WHOLE-CROP CEREALS

Second Edition

Edited by
B A Stark and J M Wilkinson

CHALCOMBE PUBLICATIONS

First published in the United Kingdom
by Chalcombe Publications,
Church Lane, Kingston, Near Canterbury,
Kent CT4 6HX

First edition 1990
Second edition 1992

© Chalcombe Publications 1992

ISBN   0 948617 25 X

Printed in the United Kingdom by Cambrian Printers, Aberystwyth

CONTENTS

FOREWORD

Since the first publication of this book, considerable interest has been shown by farmers in the concept of using the cereal crop harvested complete. The interest has been stimulated partly by restricted supplies of grass silage caused by dry summer weather. But, in addition, a strong group of farmers has developed who see the longer-term potential for whole-crop cereals. These farmers have recently been working closely with research organisations and commercial companies in an attempt to establish the value of whole-crop cereals.

We now have ample evidence to suggest that mixing forages can improve feed intake, milk yield and milk quality. Maize is now well-established as a complementary forage to grass – as indeed is alkali-treated straw. Fermented and alkaline whole-crop cereals are well-suited nutritionally to complement both grass and grass silage, and also maize.

Agronomic information indicates that, compared to grass, whole-crop cereals require lower inputs, are better suited to crop rotations, make better use of inherent soil fertility, and have lower harvesting costs. Whole-crop cereals provide a solution to the problem of straw disposal and enable home-grown grain to be used easily on the farm.

The techniques of storage and preservation of whole-crop cereals are relatively simple. There is no risk of effluent. In the case of alkaline whole-crop, aerobic stability of the conserved product is good. But we need to identify the correct stage of harvest, the appropriate additive, and minimal application rates for good preservation.

The growing of whole-crop cereals gives the livestock farmer the opportunity to grow crops which already have ample technical support. Crop yields are reliable, there is good ground cover through the winter, and harvest is in one single operation. But this forage has still to be developed through a coordinated research effort, so that our knowledge of preservation and feeding matches our ability to grow the crop. The Maize Growers' Association is now working with all sections of the industry to achieve these aims.

Colin Wright
Chairman, Whole-Crop Cereals Group
Maize Growers' Association
December 1991

PREFACE

The chapters in this Second Edition of WHOLE-CROP CEREALS are the proceedings of two meetings held in 1990 and 1991 to review the state of the art on the production and utilisation of whole cereal crops.

The first meeting was held at the AFRC Institute for Grassland and Animal Production, Hurley, now the Institute of Grassland and Environmental Research, on 17 January 1990. The proceedings of this seminar, which formed the first edition of WHOLE-CROP CEREALS, have been revised and updated for the Second Edition.

The second meeting followed the establishment later in 1990 of a Whole-Crop Cereals Group within the UK Maize Growers' Association. It was held at the Royal Agricultural College, Cirencester on 18 April 1991.

Barbara Stark
Mike Wilkinson
December 1991

CHAPTER 1

A DECADE OF RESEARCH INTO WHOLE-CROP CEREALS AT HURLEY

R M Tetlow[1]

AFRC Institute of Grassland and Environmental Research,
Hurley,
Maidenhead, Berks, SL6 5LR

SUMMARY

Agronomists at the then Grassland Research Institute began monitoring the yields, dry matter (DM) contents and chemical composition of a variety of small-grained cereal crops over 20 years ago, and this continued until 1973. Dry matter yields of winter varieties of oats and wheat, harvested between the end of May and the end of July, reached a maximum of approximately 12 tonnes DM per hectare at the end of July, at a mean DM content of 43%. Spring varieties of barley, wheat and oats, harvested from mid-June to mid-August, had a maximum yield of 10 to 12 tonnes DM per hectare in early August, at a mean DM content of 44%. In 1989 a similar study was carried out with a number of newer winter varieties of wheat, barley, oats and triticale grown on plots. Maximum yields, generally in mid-June, were 15, 11.9, 15.2 and 18.1 tonnes DM per hectare respectively. As the crops matured, their DM content increased and there were also changes in the major chemical components, and in the digestibility of the organic matter (OMD). The extent and type of fermentation when whole-crop barley and wheat were ensiled and the composition of the resultant silages depended on the DM and the water-soluble carbohydrate (WSC) contents of the crop. Losses during ensiling tended to decrease with more mature crops. The main disadvantages of the whole-crop cereal silages traditionally made in clamps at Hurley were the need to harvest at a DM content of 30 to 35%, which was reached at 80 to 90% of maximum crop yield, poor aerobic stability of the silo face, and the production of an acidic feed with a low protein content. Ten years

---

[1]Present address: 1 Woodstock Close, Maidenhead, Berks, SL6 7JT

ago a research programme was started on an entirely new approach to the storage and feeding of whole-crop cereals to try to overcome some of these disadvantages. It was decided to maximise the crop yield, thus cutting a more mature crop with a DM content of around 50%. To overcome problems of decreased aerobic stability and a more lignified straw component, in particular, treatment of various crops with sodium hydroxide (NaOH), aqueous and anhydrous ammonia (NH₃) and urea at different rates was carried out. As the maturity of harvested barley and wheat increased, the improvement in digestibility following NaOH or NH₃ treatment increased markedly. Both of these alkalis are, however, hazardous and NH₃ is difficult to apply. Urea treatment, whilst having less effect on silage digestibility, is a safer procedure and this, together with the results of experiments comparing data obtained in vitro and in vivo, led to the conclusion that the most appropriate treatment for whole-crop cereal silage is urea applied at 4% of the crop DM. As with NaOH, urea treatment results in a relatively alkaline feed, but in contrast to NaOH treatment the resultant silage has a high nitrogen content and is stable when exposed to air.

YIELDS OF WHOLE-CROP CEREALS

Research into whole-crop cereals began at what was then the Grassland Research Institute (GRI) over twenty years ago. A team of agronomists began monitoring a variety of small-grained cereal crops for their yield, dry matter (DM) content and chemical composition. This project lasted for a number of years and culminated in a technical report, published in 1977, *Whole Crop Forages. Relationship between stage of growth, yield and forage quality in small-grain cereals and maize* (GRI Technical Report No. 22, by A J Corrall, A J Heard, J S Fenlon, Cora P Terry and G C Lewis). Figures 1.1 and 1.2, respectively, present a summary of their data for yields of winter and spring varieties of different crop species.

Data in Figure 1.1 show the DM yields (tonnes per hectare) of winter varieties of oats and wheat harvested at approximately three weekly intervals between the end of May and the end of July. Maximum yield (approximately 12 tonnes DM per hectare) occurred at the end of July, at an average DM content of 43%.

Figure 1.1 Dry matter yields of winter cereals

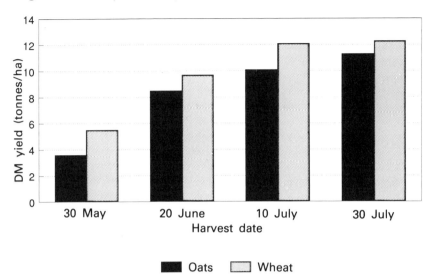

Source: Corrall *et al* (1977). GRI Technical Report No 22.

Figure 1.2 Dry matter yields of spring cereals

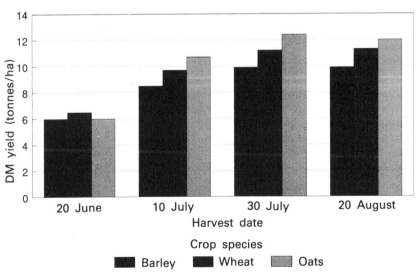

Source: Corrall *et al* (1977). GRI Technical Report No 22.

Table 1.1 Species and varieties selected for 1989 plot experiment

Cereal	NIAB recommendation
Winter wheat	
Brock	Fully recommended
Mercia	Fully recommended
Parade	Provisionally recommended
Winter barley	
Marinka	Fully recommended
Magie	Fully recommended
Frolic	Provisionally recommended
Winter oats	
Pennal	Fully recommended
Pennarth	Fully recommended
Solva	Provisionally recommended
Winter triticale	
Lasko	No recommendation

Data in Figure 1.2 show the DM yields of spring varieties of barley, wheat and oats between the third week of June and the third week of August. Maximum yield (between 10 and 12 tonnes DM per hectare) occurred in the first two weeks of August, at an average DM content of 44%.

Since those early days many new varieties have become available, and it was decided, in 1989, to examine some of these in an experiment at Hurley. Four crop species were chosen, viz. winter varieties of wheat, barley, oats and triticale. Three varieties of each of wheat, barley and oats were selected for the trial on the basis of disease resistance and stiffness of straw. One variety of triticale was studied. Table 1.1 shows the varieties chosen. Crop DM yields and dates of cut are shown in Figure 1.3.

Maximum yields for most of the plots occurred at the beginning of the third week of June. At this date crop yields for the wheat, barley, oats and triticale were 15, 11.9, 15.2 and 18.1 tonnes DM per hectare, respectively. The DM contents of the crops are presented in Figure 1.4.

Figure 1.3 Mean dry matter yields of winter cereals, plot data 1989

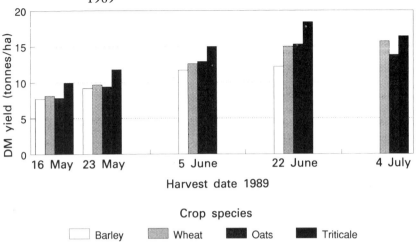

Figure 1.4 Mean dry matter contents of winter cereals harvested between 16 May and 4 July 1989

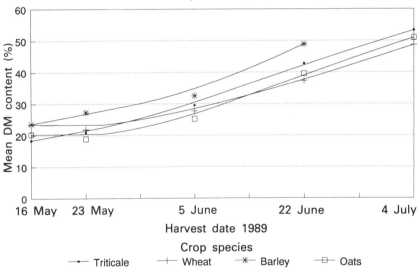

Sampling of the plots ceased when a DM content of 50% was reached. For the barley, this occurred about a fortnight before the other crops. The rate of increase of DM content was 0.67 percentage units per day. Assuming the "window" for harvesting is between 40 and 50% DM for Hurley, in 1989 this would have given a "window" of 15 days in which to complete the harvest.

Figure 1.5   Effect of crop maturity on the chemical composition of wheat

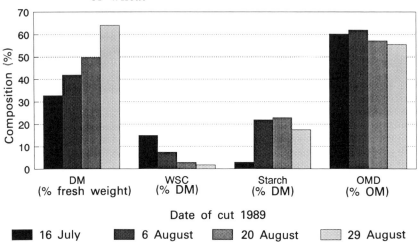

Figure 1.6   Effect of crop maturity on the chemical composition of barley

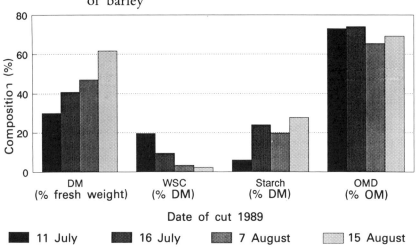

COMPOSITION OF WHOLE-CROP CEREALS

As crops mature, not only does DM content increase but there are changes in the major crop components. Water-soluble carbohydrates (WSC) decrease as they are converted to starch, which increases as the grain forms. Of particular interest is the relatively slow decline in the digestibility of the organic matter (digestible organic matter expressed as a % of the organic matter; OMD). These changes are common to all small-grained, whole-crop cereals and are demonstrated in Figure 1.5 for wheat and in Figure 1.6 for barley.

THE ENSILAGE OF WHOLE-CROP CEREALS

Fermentation

The extent and type of fermentation when whole-crop cereals are ensiled is dependent on the DM content and the level of fermentable substrate, i.e. the WSC content. The effect of increasing maturity on the fermentation of barley and wheat silages is demonstrated in Figures 1.7 and 1.8 respectively. Of note is the similarity between

Figure 1.7 Effects of crop maturity on the fermentation of whole-crop barley silage

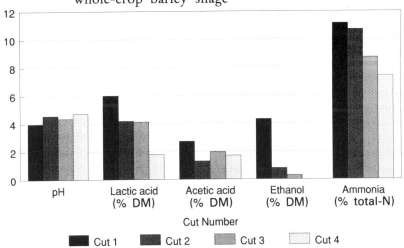

Figure 1.8 Effects of crop maturity on the fermentation of whole-crop wheat silage

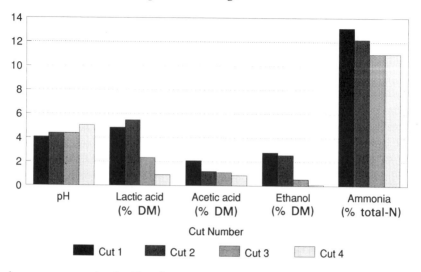

the two crops. As the DM increases and the WSC falls, the pH rises; lactic and acetic acid contents fall, as does that of ethanol. Ammonia expressed as a proportion of the total nitrogen also falls, indicating a reduction in the breakdown of protein.

Dry matter losses

Of importance during ensiling are the losses of DM which occur. These are reported in Figure 1.9. Data presented are based on samples which were oven-dried and therefore do not take into account losses of volatile components such as acetic acid, ethanol, ammonia and, to some extent, lactic acid. Nevertheless, the trend with both crops is a decrease in losses as they mature, with losses from barley at a slightly lower level compared to wheat.

Making and feeding whole-crop silages

Whole-crop silage can give good animal performance under ideal conditions. Good consolidation is essential during ensiling if losses during feedout are to be at an acceptable level. Tower silos are better than bunkers, but if bunkers are the only option then it is necessary to harvest at 30 to 35% DM; at Hurley this means cutting at between

Figure 1.9 Dry matter losses during the fermentation of whole-crop wheat and barley. Data based on oven-dried samples

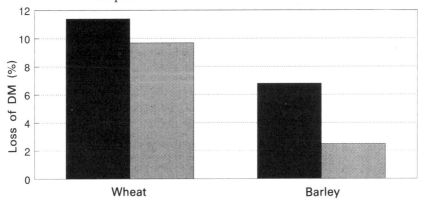

80 and 90% of maximum yield. In addition, with bunker silos feedout rate should be rapid and good silo face management is essential to ensure low feedout losses.

Nitrogen content is low so protein supplementation is necessary for young and high-producing animals. Even under ideal conditions, whole-crop cereal silages can be very acidic, with a pH as low as 3.6, and this may give rise to palatability problems or even acidosis in cattle consuming diets containing high proportions of such materials.

A NEW APPROACH TO WHOLE-CROP SILAGE

A decade ago a small team at the GRI (now the AFRC Institute of Grassland and Environmental Research, IGER) began working on an entirely new approach to storing and feeding whole-crop cereals, in an attempt to overcome some of the problems outlined above. It was decided to maximise the DM yield of the crop, when the DM content was about 50%. This led to increased problems of silo consolidation, and the straw component was highly lignified and thus less digestible.

ALKALI TREATMENT OF WHOLE-CROP SILAGES

Drawing on experience of treatment of straw with sodium hydroxide (NaOH), a programme of work was decided using alkalis to try to solve these new problems. It was believed that NaOH would alleviate the problems of consolidation and would increase the digestibility of the straw, but its preservative properties were less well known.

Data in Figures 1.10 and 1.11 respectively show the interactive effects of crop maturity and NaOH, applied to barley and wheat at 5% of the crop DM, on the digestibility of the organic matter, measured *in vitro*.

In Figure 1.10, as barley became more mature the effect of treatment with NaOH increased, with the result that at Cut 4 (DM=60%) silage treated with NaOH was over 80% digestible, 20 percentage units higher than the control silage. Data in Figure 1.11 for wheat are even more dramatic, the digestibility of NaOH-treated silages exceeding that of the control silages by more than 25 percentage units.

Figure 1.10   Effect of sodium hydroxide on the digestibility of organic matter *in vitro* of barley of increasing maturity

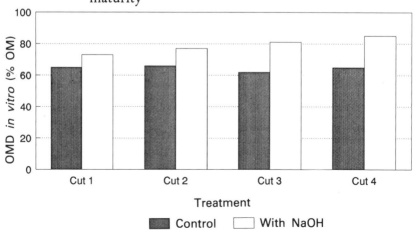

10

Figure 1.11 Effect of sodium hydroxide on the digestibility of organic matter *in vitro* of wheat of increasing maturity

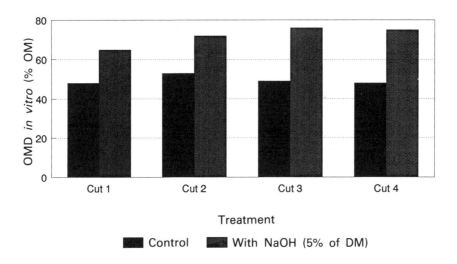

Even though these increases were dramatic, the dangerous nature of NaOH was appreciated. Consequently, an alternative alkali was sought. Ammonia (NH₃), in its aqueous and anhydrous forms, was

Figure 1.12 Effect of various alkalis on the digestibility of organic matter (OMD) of wheat *in vitro*

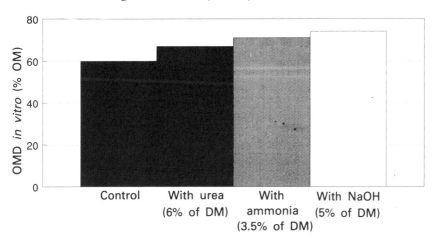

tried and found to be effective but difficult to apply. Urea was also tested and, as a result of urease enzyme in the crop, NH_3 was released. A comparison of the effects of the various alkalis on wheat is presented in Figure 1.12.

All alkalis were effective compared with the control. Although urea was the least effective, this method of introducing the alkali was very attractive and the research moved on to testing this approach against the use of NaOH. A number of experiments have been completed and the results from a study in which urea was compared directly with NaOH at a range of application rates are presented in Figures 1.13 and 1.14.

Figure 1.13 The effect of urea and sodium hydroxide applied at different rates (% of crop DM) on the digestibility of the organic matter *in vitro* of wheat silage

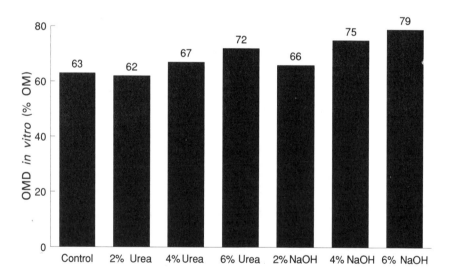

Increasing the application rates of both urea and NaOH resulted in increases in the OMD of the silages. The stepwise improvements were progressively greater for NaOH than urea, with the highest rate of 6% resulting in an OMD of 79%, which was in line with previous results. However, when these silages were given to cattle, digestibility measured *in vivo* told a different story. Comparisons in Figure 1.14 between results obtained *in vitro* and *in vivo* showed that for control and urea-treated silages there was no major discrepancy between the two, but that for NaOH there was a discrepancy which became progressively greater as the application rate increased. On the basis of these data and of those presented in Chapter 6 for intake by cattle, it was concluded that the most appropriate treatment was the application of urea at 4% of the crop DM.

Figure 1.14 Comparison of the digestibility of organic matter determined *in vitro* and *in vivo* with steers given wheat silage and soyabean meal

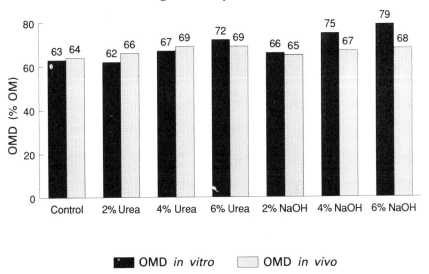

AEROBIC STABILITY OF ALKALI-TREATED WHOLE-CROP SILAGES

The effect of treatment with NaOH on the aerobic stability of barley silage is shown in Figure 1.15. The presence of NaOH conferred some control on the heating of both low and high DM silages. These effects were echoed in the loss of DM during this aerobic phase, as shown by data presented in Figure 1.16.

Figure 1.15 Aerobic stability of whole-crop barley silage

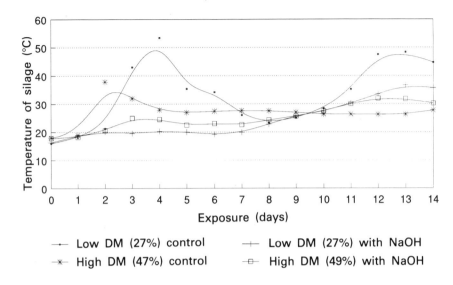

Low DM (27%) control   Low DM (27%) with NaOH
High DM (47%) control   High DM (49%) with NaOH

Figure 1.16 Aerobic losses from whole-crop barley silage

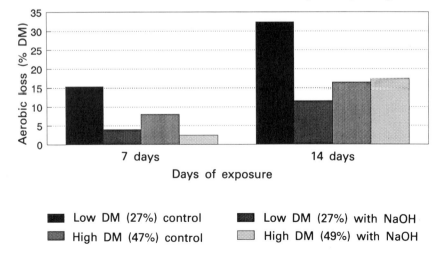

Low DM (27%) control   Low DM (27%) with NaOH
High DM (47%) control   High DM (49%) with NaOH

Figure 1.17   Aerobic stability of whole-crop wheat silage

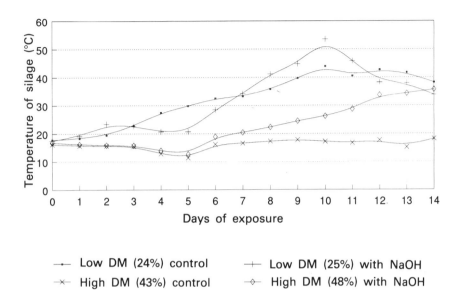

Low DM (24%) control   Low DM (25%) with NaOH
High DM (43%) control   High DM (48%) with NaOH

Figure 1.18   Aerobic losses from wheat silage

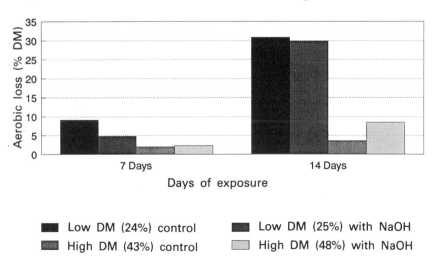

Low DM (24%) control   Low DM (25%) with NaOH
High DM (43%) control   High DM (48%) with NaOH

Data for wheat in Figure 1.17 suggest that NaOH has little or no effect on heating and on DM losses, and at high DM contents may be detrimental (Figure 1.18).

The use of urea as an additive is shown in Figure 1.19 to be highly effective in preventing heating for up to 20 days in wheat silage of high DM content. There was a temperature increase after this time, but in practice any exposed silo face would have been replaced by then.

Figure 1.19 Aerobic stability of wheat silage with a dry matter content of 55%, and treated with urea

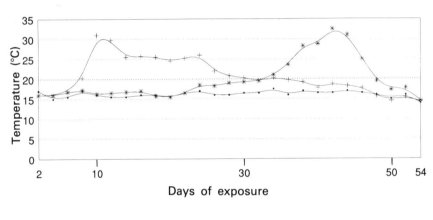

— • — Ambient temperature   —⊢— Untreated control silage
—✳— Silage treated with urea (4% of DM)

CONCLUSIONS

*   Harvest at maximum yield;
*   Ensile with urea at 4% of the crop DM.

This recommendation results in:
*   High crop yield;
*   High digestibility;
*   High nitrogen content;

* High alkalinity;
* High stability during feedout.

DISCUSSION

The time when the crop reaches 50% DM relative to the date at which the conventional point of harvest is reached is variable, depending on factors such as weather, soil type and cereal variety. In a very hot, dry year such as 1989, 50% DM was reached at an atypical time. Experience at IGER suggests that the crop will have a DM content of about 50% when the grain is at a medium/hard dough stage, the standing crop looks neither uniformly green nor pale yellow and, at least with wheat, there is approximately 2.5cm (1 inch) of green either side of the second node from the base of the plant.

With alkali-treated whole-crops the aim is to harvest between 45% and 55% DM content, which gives a relatively narrow time "window". Wetter crops than this are likely to ferment and, as the addition of alkalis (sodium hydroxide, ammonia or urea) buffers changes in pH and prevents an adequate level of acidity being reached, a clostridial fermentation may result. At IGER, butyric acid levels in silages made from wet crops have reached 15% of the DM of whole-crop cereals treated with sodium hydroxide. Conversely, when crops are harvested with DM contents above about 75% there may be inadequate moisture present to allow urea hydrolysis, and the significant amounts of urea remaining in the feed and entering the rumen may cause ammonia toxicity.

In Scandinavia whole-crop cereals are generally harvested at an earlier stage, about 5 weeks after the start of heading. Although the digestibility of the resulting material is higher than with later-cut crops, there are problems of aerobic stability. The two main nutritional differences between early and late cut cereals is that the straw of earlier-cut crops is less lignified and therefore more digestible, and in the later-cut cereals more sugars have been converted to starch. There is, however, no evidence to suggest that the presence of starch rather than sugar is nutritionally advantageous.

The addition of urea increases the aerobic stability of the crop. There are, however, other methods of preventing deterioration of stored material, such as the Ag-Bag system. Here less mature crops, with a lower DM, are ensiled in a sealed polythene tube in the absence of air, to produce a fermented material. Relatively small quantities of silage are exposed to air at feed-out.

The crop yields presented relate to cereals grown on plots. However, some data are available for field harvested crop yields at IGER, when silage has been made for feeding trials with animals. Yields from cereals grown on plots and harvested crop yields may vary quite markedly due to field losses, but at IGER harvested yields in excess of 12 tonnes DM per hectare were regularly obtained. Even with less mature crops harvested at 30% to 35% DM content, yields were relatively high.

Urease, which is the enzyme catalysing the conversion of urea to ammonia, is of microbial rather than plant origin. The moisture content of the harvested material is relatively more important than the ambient temperature, as bacteria require a moisture level of at least 22% to grow; during the harvest period ambient temperatures are normally relatively high (around 20°C), and adequate for microbial growth. At IGER the whole-crop was consolidated so that the silo held as much material as possible and, providing that the moisture content was adequate, problems of lack of urea hydrolysis to ammonia did not arise.

To date, research at IGER has covered several cereal species. Agronomic conditions for whole-crop cereals have been similar to those used for growing cereals for conventional harvesting. Specific cultural techniques for whole-crops have not been studied, but may be significant. For example, even chemically or biologically treated straw has a much lower energy content than the grain fraction, thus the use of a straw shortener is important. The conventional use of agrochemicals does not cause difficulties, despite the earlier harvesting date, providing that the instructions regarding timing for each product are followed. With the Ag-Bag system applications of nitrogen fertiliser tend to be increased by about 10% but, because the crop is harvested earlier, lodging is not a problem. In addition, there is usually omission of the last fungicide application and that for ear emergence.

The ratio of grain:straw in the whole-crop has been measured on occasions at IGER, with results varying from 40 to 60% grain.

Although urea has an effect on increasing the digestibility of the grain husk, this is relatively less important than its effects on the straw. Discrepancies may occur between results of laboratory studies and *in vivo* trials due to grinding the husk for the former. There appears to be good agreement between the results of laboratory studies, animal trials and farming experience for the cereals studied at IGER, except barley. Reasons for discrepancies between recommendations based on laboratory studies with barley and results from animal trials, for example in optimum DM content at harvest, are not clear.

CHAPTER 2

THE PRODUCTION AND FEEDING OF WHOLE-CROP CEREALS AND LEGUMES IN DENMARK

V F Kristensen

National Institute of Animal Science, Foulum, P O Box 39, DK-8830 Tjele, Denmark

SUMMARY

Whole-crop silage has become a popular and widely used cattle feed in Denmark. Barley has until now been the most important crop for this purpose, although mixtures of barley and peas are also common. Pure legume crops (peas and field beans) have only been ensiled to a limited extent. The use of winter wheat is increasing. The yield potential of winter wheat is 9 to 17 tonnes of dry matter (DM) per hectare, of spring barley 6 to 11 tonnes and of peas or field beans 6 to 9 tonnes. Spring crops are normally undersown and 2 to 2.5 tonnes of grass is grazed or harvested in the autumn. The metabolisable energy (ME) value of whole-crop cereals is 9.4 to 10.7 (average 10.0) MJ per kg DM, that of field beans about 10.5 MJ and that of peas about 11.5 MJ per kg DM. The optimum harvesting time based on studies of yield, composition and feeding value was found to be 5 to 6 weeks after the start of heading of winter wheat, with a DM content around 40%, and for barley 4 to 5 weeks after the start of heading, with a DM content around 35%. Peas and field beans are harvested when the pods are well-filled. Whole-crop cereals and legumes of 30 to 50% DM are easy to ensile without additives. The aerobic stability of cereal silages is, on the other hand, relatively poor. This problem is in practice overcome by effective packing and consolidation, a reasonable relationship between surface area and the daily amount of silage used, and a sharply cut silage face. Whole-crop silages are mainly used as a feed for dairy cows. A great variety of supplementary feeds may be used with them. The milk production of dairy cows fed whole-crop silages does not differ from that of cows fed a high quality grass silage, but the protein content of the milk may be slightly higher.

INTRODUCTION

Research on the production and conservation of whole-crop cereals in Denmark began about 1970 in the Department of Forage Crops, Danish Research Service for Plant and Soil Science. It was soon followed by feeding experiments at the National Institute of Animal Science. Based on these studies, a system of harvesting and ensiling whole-crop cereals has been developed, and whole-crop silage has become an important part of the conserved forage on many dairy farms. At first it gained popularity as a buffer crop. The decision to harvest and ensile a whole-crop cereal could be made late in the growing season, if the yield of other forages was inadequate. It could also generally increase the forage production on farms with a high stocking rate. More latterly, whole-crop silage has proved to be more competitive than other forage crops under Danish conditions, because of comparatively low costs of harvesting and fertilising, and because it produces no effluent and consequently results in few environmental problems.

According to public statistical information (Danmarks Statistik), whole-crop cereals have been harvested on 50,000 to 60,000 hectares each year during the 1980s. The average yield, according to the same source, has been approximately 90,000 MJ ME per hectare. Total production is about 7,000 MJ ME per cow per year for the total dairy cow population.

Spring barley has until now been the most important crop for this purpose. Mixtures of barley and peas are used to some extent, and interest in the use of winter wheat is growing and expected to increase in the coming years. Pure crops of legumes (peas and field beans) have been used to a limited extent. Precise data for single crops do not exist.

The feeding values of whole-crop oats and rye are not satisfactory. Small amounts of these crops, especially oats, are harvested as green forage.

COMPOSITION, FEEDING VALUE AND YIELD OF WHOLE-CROPS

It must be emphasised that large differences exist in the composition and feeding value of cereal whole-crops. This variation is mostly due to differences in the proportion of grain and straw between crops. It is, therefore, dependent on varietal differences in length of

Table 2.1 Digestibility of organic matter *in vitro* (DOM) and DM content of farm samples of whole-crop silages

Crop	Year	No. of samples	DOM (%)	SD[1]	DM (%)	SD[1]
Barley	1983	274	72	5	37.9	9.0
	1984	490	69	4	33.2	6.8
	1985	328	68	4	32.6	4.9
	1986	863	70	5	35.7	6.6
	1987	2600	67	4	30.0	5.5
	1988	2839	70	4	33.4	6.4
	1989	3031	73	4	40.5	7.7
	1990	2950	71	3	36.1	7.7
Barley + peas	1984	124	69	4	30.9	7.0
	1985	68	68	4	31.5	7.4
	1986	81	72	4	36.0	8.4
	1987	242	69	4	27.9	4.8
	1988	201	71	4	31.5	7.0
	1989	274	75	3	40.1	9.1
	1990	527	73	3	38.1	9.7
Wheat	1986	21	69	3	37.0	5.8
	1987	100	65	5	36.8	9.1
	1988	69	71	5	42.0	7.6
	1989	189	72	4	47.1	8.5
	1990	400	70	4	41.1	8.3

[1] Standard deviation.

Source: Danish Agricultural Advisory Centre.

straw, while differences between years are caused by differences in growing conditions influencing the height of the crop and the development of the grains. Some variation may be due to the stage of development of the crop at the time of harvest, as discussed later.

Table 2.1 shows the results of analyses of samples sent by farmers for determination of the feeding value of whole-crop silages. Only results from determinations of the digestibility of the organic matter (DOM) *in vitro* are given. The DOM (% OM) values of various species have been found to be similar, although mixed cultures of barley and peas have tended to give slightly higher values than those for pure barley and wheat in recent years. The average value per year has varied from 67% to 73% for barley, from 68% to 75% for barley plus peas and from 65% to 72% for wheat. Relatively low values were obtained in 1987, when the summer was very wet and cold. The overall average value of DOM was 70%, with most of the yearly averages between 68% and 72%, resulting in a variation in the ME value from 9.7 to 10.2 MJ per kg DM. The standard deviation for DOM within years was about 4 units.

The average yearly DM content ranged from 30% to 40% for barley silage, from 28% to 40% for mixed cultures of barley and peas, and from 37% to 47% for wheat, with the standard deviation within years being about 7 units. This standard deviation indicates a great variation in the stage of development of the crops at harvest, which may explain part of the variation in feeding value. In addition, difficulties in obtaining representative samples of whole-crops could account for part of the variation in the feeding value of the samples.

Field beans and peas have generally high feeding values, field bean silage being about 10.5 MJ ME per kg DM. The DOM (% OM) measured in sheep is about 75% for field beans and about 80% for pea silage.

The yield of DM per hectare obtained in experiments has been 8 to 12 tonnes for spring barley, 9 to 17 tonnes for winter wheat, 7 to 10 tonnes for peas, and 7 to 12 tonnes for field beans.

Optimum harvesting time

Figure 2.1 shows the crop yield in tonnes of DM per hectare for

Figure 2.1 Dry matter yield of whole-crop winter wheat in relation to heading and the time of harvesting

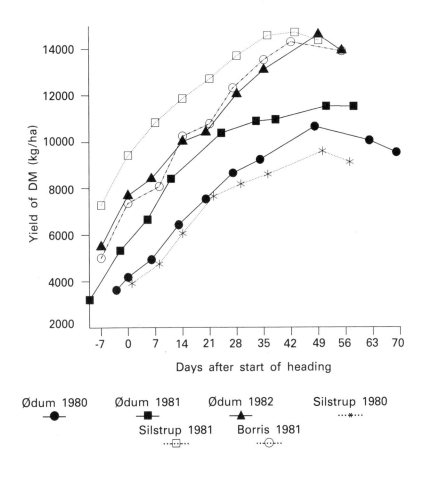

6 experiments. The crop was fertilised with 3 levels of nitrogen (N) and the results shown are mean values for the N levels. The wheat crop was harvested at frequent intervals from about, or a few days before, initial ear emergence until 7 to 10 weeks later. The experiment was carried out at 3 experimental stations (Ødum, Silstrup and Borris) during 3 years.

The maximum yield of DM per hectare was, in all cases, obtained

Figure 2.2 Digestibility of organic matter *in vitro* of winter wheat in relation to heading and the time of harvesting

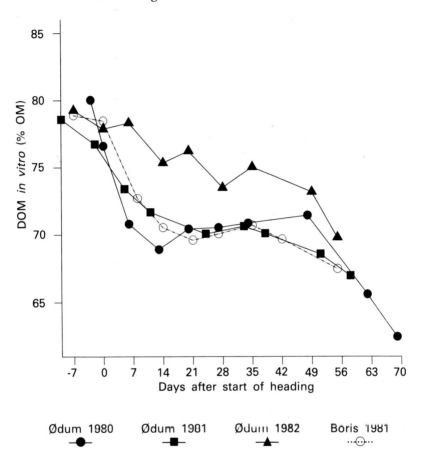

5 to 7 weeks after initial ear emergence. The DOM (% OM) usually dropped sharply from the time of initial ear emergence until the end of heading (Figure 2.2). During the period from approximately 2 weeks to 7 to 10 weeks after initial ear emergence, the DOM tended to increase slightly. After this, it dropped again. Maximum yield per hectare of digestible OM was therefore also obtained 7 to 10 weeks after initial ear emergence.

The DM content of the crops increased progressively with increas-

Figure 2.3 Dry matter content of winter wheat in relation to the time of harvesting

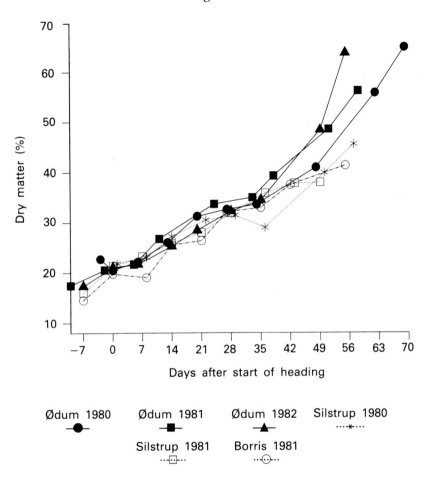

ing growth stage (Figure 2.3). Six weeks after the start of heading it was 35 to 43%. The protein content (N x 6.25) dropped during the period from before heading to approximately 5 weeks after initial ear emergence (Figure 2.4). In two experiments with low crop yields per hectare, the protein content levelled off at about 10% of the DM. In other cases it fell to 6 to 8% of the DM.

The content of water soluble carbohydrates (WSC) was measured in 3 experiments. It remained high (14 to 25% of the DM) until 4 to

Figure 2.4 Crude protein content of winter wheat in relation to heading and the time of harvesting

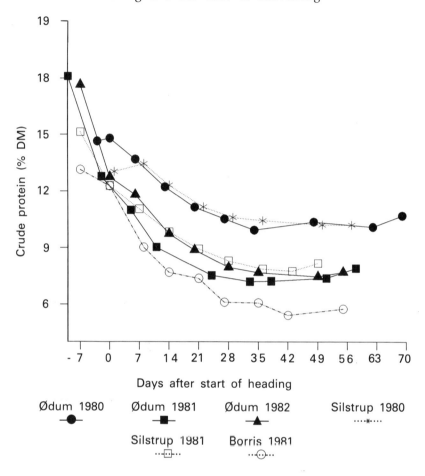

Ødum 1980
●

Ødum 1981
■

Ødum 1982
▲

Silstrup 1980
····*····

Silstrup 1981
····□····

Borris 1981
····○····

5 weeks after initial ear emergence (Figure 2.5), after which it fell sharply.

Similar changes in yield and composition in relation to stage of development occur in spring barley. However, while maximum yield per hectare in winter wheat is obtained at a DM content of about 40%, in spring barley maximum yield is obtained at approximately 35% DM of the crop.

Winter wheat harvested at 3 stages of development (approximately 3, 5 and 7 weeks after initial ear emergence) has been evaluated in

Figure 2.5 Water soluble carbohydrate content of winter wheat in relation to heading and the time of harvesting

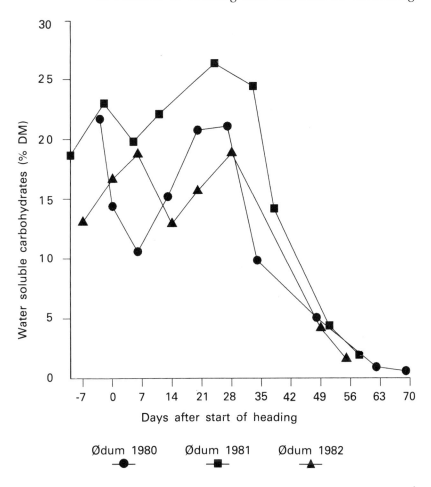

feeding experiments with dairy cows. Data for the composition of the wheat silages are shown in Table 2.2.

A short-strawed type of wheat (Slejpner) was used. The DM content increased from 30.6% to more than 50% from the first to latest harvest time. Crude protein content decreased from 10.5% to 8.4%, and neutral detergent fibre (NDF) from 42.8% to 33.3% of the DM. The starch content (DM basis) increased from 8.5% to 30.8%, and the energy value increased slightly from 10.2 to 10.8 MJ ME per kg DM.

Table 2.2 Chemical composition and feeding value of winter wheat (Slejpner) harvested 3, 5 and 7 weeks after the start of heading

| | Harvest date | | |
	5 July	19 July	2 August
DM (%)	30.6	38.5	52.4
Crude protein (% DM)	10.5	9.4	8.4
NDF (% DM)	42.8	35.7	33.3
WSC (% DM)	7.1	4.2	3.1
Starch (% DM)	8.5	21.1	30.8
DOM (%)	71	74	75
ME (MJ/kg DM)	10.2	10.6	10.8

The feeding experiment was carried out over 2 years with dairy cows in early lactation (weeks 3 to 15 after calving), and 18 cows per treatment group. Fixed daily amounts of fodderbeet (4.7 kg DM), soyabean meal with 17% added fat (3.0 kg DM), and rolled barley (1.3 kg DM) were given together with wheat silage *ad libitum*.

The intake of winter wheat silage was in all cases about 9 kg DM per cow per day and was not significantly affected by time of harvest. The fat content of the milk fell from 4.11% with early harvesting to 3.81% with the latest harvest time (Table 2.3). Milk protein content

Table 2.3 Daily milk production in relation to time of harvest of whole-crop winter wheat

Date of harvest	5 July	19 July	2 August
Milk yield (kg)	27.3	28.2	27.7
Milk fat (g/kg)	41.1	39.8	38.1
Milk protein (g/kg)	29.9	30.1	30.0
ECM* (kg)	27.1	27.7	26.6

*Energy corrected milk (3140 KJ/kg) based on fat and protein content.

was not affected. The yield of energy corrected milk was highest at the intermediate harvest time, and it corresponded to the DOM in the total diet measured in the cows, which was also highest at the intermediate harvest time. The drop in fat content was probably due to the increase in the content of starch and the decrease in the content of digestible fibre in wheat silage.

The loss of OM during ensiling in this experiment was reduced from 12 to 16% with the early harvest time to close to zero at the latest harvest. Net yield of ME was highest with the latest harvest time.

Based on the results of all experiments, it was concluded that the optimum harvest time for whole-crop wheat is 5 to 6 weeks after the start of heading, when the grains are at the soft dough stage and the DM content of the crop is around 40%.

From similar studies on whole-crop barley, it was concluded that this crop should be harvested 4 to 5 weeks after the start of heading or when the DM content is around 35%.

CARBOHYDRATE SOURCES AS SUPPLEMENTS FOR WHOLE-CROP SILAGE

Rolled barley, fodderbeet and dried sugar beet pulp were compared as supplements for whole-crop barley silage fed *ad libitum* to dairy cows from week 3 to week 20 after calving (Table 2.4), with

Table 2.4 Quantities (kg DM per cow per day) of different carbohydrate sources fed as supplements to whole-crop barley silage

| | Treatment | | | | | |
Supplement	1	2	3	4	5	6
kg DM per cow per day						
Rolled barley	5.2	2.6	–	2.6	–	–
Dried sugar beet pulp	–	–	–	2.6	2.6	5.2
Fodderbeet	–	2.6	5.2	–	2.6	–
Soyabean meal + fat	3.0	3.3	3.6	3.0	3.3	3.0

Table 2.5 Silage DM intake (kg per cow per day) with different carbohydrate sources as supplements to whole-crop silage

Supplement	Silage intake (kg DM/cow/day)
Barley	9.0
Barley/fodderbeet	9.5
Fodderbeet	8.8
Barley/sugar beet pulp	10.6
Sugar beet pulp/fodderbeet	9.1
Sugar beet pulp	10.0

12 cows per treatment group. The DOM of the barley silage was 71% and its ME value was 10.1 MJ per kg DM.

The intake of silage DM was highest with a supplement of barley plus sugar beet pulp, and with sugar beet pulp alone (Table 2.5). There were no significant differences in the daily yield of milk, milk fat or 4% fat corrected milk (FCM) (Table 2.6). The protein content and the daily yield of milk protein were highest when rolled barley was given as a supplement.

Table 2.6 Milk yield and composition with different carbohydrate sources as supplements to whole-crop barley silage

Supplement	Milk yield (kg/day)	Milk fat (g/kg)	(g/day)	Milk protein (g/kg)	(g/day)	4% FCM* yield (kg/day)
Barley	28.1	40.0	1124	31.7	891	28.1
Barley/fodderbeet	27.8	42.2	1172	31.2	866	28.7
Fodderbeet	27.2	42.1	1145	30.1	820	28.1
Barley/sugar beet pulp	27.6	41.6	1147	31.6	872	28.3
Sugar beet pulp/ fodderbeet	27.7	40.9	1134	30.1	833	28.1
Sugar beet pulp	27.3	41.9	1144	30.0	818	28.1

*Fat corrected milk.

Table 2.7 Milk yield and composition with fodderbeet compared to dried sugar beet pulp as a supplement for whole-crop winter wheat silage

Kg DM from fodderbeet	Milk yield (kg/day)	Milk fat		Milk protein		Yield of fat + protein (g/day)	ECM* yield (kg)
		(g/kg)	(g/day)	(g/kg)	(g/day)		
5.4	28.6	45.2	1287	31.7	900	2187	30.0
2.7	29.6	41.2	1218	29.7	870	2087	29.3
0	29.2	42.3	1234	29.8	870	2105	29.4

*Energy corrected milk (3140 KJ/kg) based on fat and protein content.

In another experiment, 5.4 kg of fodderbeet DM was either half or completely substituted by dried sugar beet pulp as a supplement to winter wheat silage. Fodderbeet gave higher fat and protein contents in the milk, but only small differences in the daily production of fat, protein and energy corrected milk (Table 2.7).

High levels of milk production may be obtained with diets based on whole-crop barley or wheat silage, and a wide variety of supplemental feeds may be used with only minor effects on the production of milk and milk components.

WHOLE-CROP LEGUME SILAGE

Peas and field beans are to some extent used for whole-crop silage, but a mixed crop of spring barley and peas is more common. A few results from an experimental comparison of whole-crop barley with field beans and peas are given in Tables 2.8 and 2.9. The feeding value of the barley used in the experiment was rather low, possibly because it was harvested too early. The crops were wilted before ensiling. The feeding values of the field beans and peas are typical (Table 2.8); in particular peas generally have a very low fibre content and a high digestibility and feeding value.

The experimental period lasted from week 23 to week 32 of lactation. Feed intake, milk production and liveweight change were higher with pure legume silages than with barley or a mixture of barley and field beans (Table 2.9). The highest production was

33

Table 2.8 Composition and feeding value of whole-crop barley, legumes and mixed crop silages

	Barley	Barley + field beans	Field beans	Peas
DM (%)	37.2	38.2	27.1	31.0
% in DM				
Crude protein	8.3	12.6	15.5	14.9
Crude fibre	26.4	27.0	27.2	21.0
NDF	63.2	54.7	43.7	41.5
ADF	32.5	33.0	32.9	26.1
WSC	9.2	9.3	3.0	4.7
Starch	10.8	8.0	12.9	17.6
DOM (%)	67	66	74	79
ME (MJ/kg DM)	9.5	9.3	10.5	11.3

Table 2.9 Performance of dairy cows fed whole-crop barley, legume or mixed crop silages

Silage	Silage DM intake (kg/day)	Yield of 4% FCM* milk (kg/day)	Daily weight change (kg)
Barley	9.2	23.6	0.251
Barley + field beans	8.9	23.5	0.339
Field beans	10.1	24.4	0.490
Peas	10.6	25.9	0.487

*Fat corrected milk.

obtained with peas. The differences observed in feed intake and milk production were in accordance with differences in the energy intakes of the cows.

Field beans do not grow satisfactorily in mixed cultures with barley, and relatively low yields and feeding values are obtained. On the other hand, yields and feeding values of mixtures of barley and peas are intermediate between those of pure barley and pea crops.

Field beans and peas are harvested when all pods are well filled.

COMPARISON OF WHOLE-CROP BARLEY AND CLOVER/GRASS SILAGES

Whole-crop barley silage was gradually substituted by clover/grass silage in an experiment with dairy cows (Table 2.10). The energy values of the barley and clover/grass silage were similar (10.1 MJ ME per kg DM).

The protein content of the milk and the yield of milk protein were higher with barley silage than with grass silage. The yields of 4% FCM on the barley and grass silages were not significantly different. The lower yield on 2:1 whole-crop:clover/grass silage probably occurred by chance.

Table 2.10 Milk production of dairy cows fed whole-crop barley silage in mixtures with clover/grass silage

	Yield of 4% FCM* (kg/day)	Milk protein (g/kg)	(g/day)
Whole-crop	28.6	31.6	863
Whole-crop: clover/grass (2:1)	26.6	31.2	815
Whole-crop: clover/grass (1:2)	28.3	30.9	837
Clover/grass	28.1	30.1	816

*Fat corrected milk.

CONCLUSIONS

Potential yields under Danish conditions are 9 to 17 tonnes DM per hectare of winter wheat and 7 to 12 tonnes per hectare of spring barley, mixed cultures of peas and barley, and pure legume crops. Spring sown crops are normally undersown with Italian ryegrass which is grazed or harvested during September and October, and which yields 2 to 2.5 tonnes of DM per hectare. The feeding value of whole-crop cereals may vary between 9.4 and 10.7 MJ ME per kg DM, while that of field beans is about 10.5 MJ ME and that of peas about 11.5 MJ ME per kg DM.

Whole-crop cereals and legumes with 30 to 50% DM are generally easy to ensile. A drop in the content of WSC follows an increase in the DM content. Because the requirement for WSC to obtain an efficient lactic acid fermentation is inversely proportional to the DM content, the WSC content is always high enough. Pure crops of legumes may sometimes be wilted in order to obtain 30% DM, but cereals are always direct harvested. These crops are normally ensiled in bunker silos or stacks, and the aerobic stability of cereal silages after opening the silo is relatively poor. Effective packing, a reasonable relationship between surface area and the daily amount of silage taken from the silo, and a sharply cut surface are essential conditions in order to obtain low ensiling losses.

In Denmark whole-crop silage is mainly fed to cows. Because of the relatively low ratio of N to digestible carbohydrates, cereal silages are in many cases supplemented with 10 to 20 gram urea per kg silage DM.

DISCUSSION

In Denmark whole-crop cereals are fermented rather than ensiled with urea, and Danish farmers have had no experience with alkali treatment. Consolidation in the clamp is difficult for crops above 40% DM, but is less critical if sealing is excellent. The addition of molasses can improve consolidation initially, but the stack sinks and the densities of untreated and molasses-treated clamps are similar after 2 months.

Aerobic stability of the ensiled material is poor because of the high content of sugars and WSC, as relatively little fermentation occurs.

Whole-crop silage tends to be twice as unstable as grass silage.

Urea added in trials (17 to 20 grams urea per cow per day) is to correct an imbalance between N and digestible energy supply – it is not used to treat the crops to improve preservation and digestibility. Urea is an effective source of N for whole-crops, but ongoing experiments are studying the best type of N to use. A urea solution is sprayed onto the material at feeding.

Whole-crop cereals are always direct cut in Denmark, generally with a combine harvester with a different head.

CHAPTER 3

ASSESSING WHOLE-CROP CEREAL MATURITY IN THE FIELD

J J Harvey

ADAS, Staplake Mount, Starcross, Exeter, Devon, EX6 8PE

SUMMARY

The development of winter wheat dry matter (DM) contents, and changes in morphology with advancing crop maturity were recorded in 1990 on 11 farms scattered from County Durham to Cornwall. On an ADAS variety trial site in Cornwall detailed comparisons were made of 8 wheat, 1 triticale and 8 winter oat varieties, including some measurements of the digestibility of organic matter in the DM (DOMD) in vitro. Crop colour was found to be a good indicator of whole-crop DM, and was easier to use than the Zadoks numerical key to grain maturity. Description of grain maturity by the terms 'milky', 'soft Brie', 'soft Cheddar', 'hard Cheddar' and 'too hard to penetrate with thumbnail' was found to be an easy-to-use, repeatable procedure. For wheat and triticale, a good indicator of whole-crop cereals at 50 to 60% DM was a predominantly yellow field colour with traces of green, and grain with a consistency of Cheddar cheese. The DOMD in vitro was in the range 63 to 71.4%. At up to 46% DM the crop was still predominantly green, while above 64% DM most of the grains were too hard to penetrate with a thumbnail. Oats differed from wheat and triticale in that they turned yellow and produced hard grains at lower DM contents. At around 50% DM the oat crop was yellow and the grains varied from 'hard Cheddar' to 'too hard to penetrate with a thumbnail'. The DOMD in vitro was 50.4% for Aintree and Craig varieties.

INTRODUCTION

Knowledge of the dry matter (DM) content of standing cereal crops which are to be conserved as the whole-crop is important for several reasons. The maximum yield of whole-crop wheat is at a crop DM content of about 60% (Corral, Heard, Fenlon, Terry and

39

Lewis, 1977), which often occurs 2 to 3 weeks before the crop is ready for combining, while effective preservation with urea requires the material to be between 45 and 55% DM; material which is too wet tends to ferment and to react adversely with the urea, and crops which are too dry tend to have a lower digestibility and to have insufficient moisture to hydrolyse the urea.

Whole-crop cereal samples submitted to ADAS had DM contents which ranged from 30 to 80%, therefore it appeared that farmers found difficulty in recognising crops of around 50% DM. Little technical information was available to help advise farmers, thus it was decided to carry out a survey around the country in 1990 to obtain information on changes in the morphology and appearance of different crops in relation to increases in their DM content.

OBJECTIVES OF THE SURVEY

* To provide farmers with an easy, in-field method of determining the DM content of whole-crop cereals so that they can judge when to conserve the crop.
* To record differences in the above between species, varieties and sites.
* To observe rates of crop drying.

METHODS

On 11 farms, widely scattered from County Durham to Cornwall, weekly observations were made of whole-crop wheat fields destined for conservation with urea, and samples were taken to determine their DM content. Each field was sampled in 5 places per visit, and on 4 farms yield assessments were carried out from quadrat cuts at each sampling date. The range in crop DM covered was 30 to 80%.

In Cornwall, ADAS carried out detailed weekly assessments on 8 wheat, 1 triticale and 8 oat varieties, all grown in variety trials in the same field. The organic matter digestibility as a percentage of the DM (DOMD) was determined on some varieties, at a whole-crop DM content of about 50%.

CROP APPEARANCE IN RELATION TO DRY MATTER CONTENT

Table 3.1 (pages 42 and 43) was constructed to summarise the results of the ADAS survey.

Having prepared this table of observations, descriptions were written of the crops at particular DM contents. The validity of these descriptions was then checked against the 75 descriptions made by the 11 farmer co-operators. There was found to be complete uniformity between ADAS and farmer recorders in the relationships observed between crop colour and DM content.

Grain textures assessed by the Zadoks numerical key, however, correlated poorly with the DM of the crop and a wide range of grain maturities were present at any particular whole-crop DM content. Grain texture assessed by verbal description as 'milky', 'soft Brie', 'soft/hard Cheddar' and 'too hard to penetrate with thumbnail' correlated better with crop DM content and these terms were easier to use than the key by Zadoks, Chang and Konzak (1974).

Changes in the wheat and triticale varieties studied

Crop DM 32 to 35%: the overall colour of the whole field is green, and the ears are entirely green. The grains contain milk but they are predominantly at the soft, unripe Brie cheese stage (not runny Brie). This corresponds to grain growth stages in the range 71 to 83, or the late milk to early dough stage.

Crop DM 36 to 38%: the whole field and the ears still appear green, and the grain is at the soft Brie stage; grain growth stages are 78 to 83.

Crop DM 39 to 42%: the field is still green overall, but the ears are turning yellow. The grain is still soft Brie consistency, at growth stages 75 to 85. Fingernail impressions are not held by the grain.

Crop DM 43 to 46%: the field is greeny-yellow, with green still the dominant colour, and the cereal ears are yellow. The grains are the consistency of soft Cheddar cheese and are still quite 'plastic' so that fingernail impressions are generally not held; growth stages are 83 to 87.

Crop DM 47 to 54%: the overall field colour is yellow, or yellow with a hint of green, while the ears are yellow or blanching. No milk

Table 3.1 Summary of survey data relating to the dry matter content, crop colour, grain maturity and DOMD *in vitro* of 8 varieties of wheat and 1 triticale (Lasko) observed on 5 dates

Variety	DM (%)	Whole plot colour	Ear colour[1]	Grain texture[2]	Zadoks growth stage	In vitro DOMD (%)
26 June						
Brock	34.5	G	G	SB	77-83	–
Slejpner	32.0	G	G	SB	77-83	–
Mercia	33.0	G	G	SB	77-83	–
Apollo	35.8	G	G	SB	77-83	–
Dean	33.9	G	G	SB	77-83	–
Beaver	32.0	G	G	SB	77-80	–
Haven	32.4	G	G	SB	77-80	–
Galahad	32.9	G	G	SB	77-83	–
Lasko	34.1	G	G	SB	77-83	–
4 July						
Brock	37.0	G	G	SB	83	–
Slejpner	34.3	G	G	SB	83	–
Mercia	35.4	G	G	SB	83	–
Apollo	37.4	G	G	SB	83-85	–
Dean	35.3	G	G	SB	83	–
Beaver	33.2	G	G	SB	83	–
Haven	34.0	G	G	SB	83	–
Galahad	35.0	G	G	SB	83	–
Lasko	35.1	G	G	SB	83	–
12 July						
Brock	42.2	G	GY	SB	83-85	–
Slejpner	39.5	G	GY	SB	83-85	–
Mercia	40.3	G	GY	SB	83-85	–
Apollo	43.3	G	GY	SB	83-85	–
Dean	42.3	G	GY	SB	83-85	–

Variety	DM (%)	Whole plot colour	Ear colour[1]	Grain texture[2]	Zadoks growth stage	In vitro DOMD (%)
Beaver	38.7	G	GY	SB	83-85	–
Haven	39.1	G	GY	SB	83-85	–
Galahad	40.1	G	GY	SB	83-85	–
Lasko	39.2	G	GY	SB	85	–
19 July						
Brock	44.5	GY	Y	SC	85	–
Slejpner	43.3	GY	Y	SC	85	–
Mercia	43.4	GY	Y	SC	85	–
Apollo	48.0	GY	Y	SC	85	–
Dean	45.5	GY	Y	SC	85	–
Beaver	41.8	GY	Y	SC	85	–
Haven	41.7	GY	Y	SC	85	–
Galahad	42.8	GY	Y	SC	85	–
Lasko	43.6	Y	Y	SC	85	–
27 July						
Brock	48.1	YG	Y	HC	87-91	–
Slejpner	48.6	YG	Y	HC	87-91	71.4
Mercia	46.8	YG	Y	HC	87-91	67.1
Apollo	54.1	Y	Y	HC	91-92	63.7
Dean	51.0	Y	Y	HC	87-91	–
Beaver	45.0	Y	Y	SC-HC	85-87	–
Haven	46.0	Y	Y	SC-HC	87-92	–
Galahad	50.3	Y	Y	SC-HC	87-92	67.8
Lasko	51.6	Y	Y	HC	87-95	63.11

[1] G = green; GY = greeny yellow; Y = yellow.
[2] SB = soft Brie; SC = soft Cheddar; HC = hard Cheddar.

Source: ADAS, Cornwall.

can be squeezed out of the grains, which are predominantly hard Cheddar consistency. Thumbnail impressions are held and some grains are too hard to penetrate with a thumbnail. The growth stages are 85 to 92 and DOMD *in vitro* ranged from 63.1 to 71.4%.

Crop DM 55 to 65%: the overall crop colour is yellow (bleached) or yellow/brown, i.e. harvest-ripe colour but with hints of green on the stem and ear. Grains are like hard Cheddar, with some too hard to penetrate with a thumbnail, and they are at growth stages 87 to 92.

Crop DM 66 to 70%: the crop colour is bleached yellow/brown, harvest-ripe colour but with traces of green at the nodes. Grains are hard, at a growth stage 92 to 93, and it is difficult or impossible to penetrate the grain with a thumbnail.

Crop DM 75 to 80%: the crop colour is yellow (harvest-ripe colour). There are, however, still traces of green at the first node. The grains are too hard to penetrate with a thumbnail and they loosen in daytime; the growth stage is 93.

Assessing the dry matter content of wheat

Taking the optimum harvest window of whole-crop cereals as 50 to 60% DM, for farmers a good indication of this range is that the field colour is predominantly yellow and the grain is the consistency of Cheddar cheese. Up to 46% DM, the field is still predominantly green. At over 65% DM, most of the grains are too hard to penetrate with a thumbnail. Table 3.2 provides the basic facts a farmer needs to know to judge when to harvest wheat for treatment with urea.

Changes in the oat varieties studied

The oat varieties Lustre, Pennal, Peniarth, Kynon, Aintree, Solva, Image and Craig were assessed in a similar way to the wheats and triticale in an adjacent trial:

Crop DM 25 to 35%: the field is green, and the grain is a soft Brie texture; grain growth stages are 77 to 83.

Crop DM 36 to 43%: the field is predominantly yellow, but still with some green in the stems. The grains are soft to hard Cheddar, and the growth stages are 83 to 92.

Crop DM 44 to 47%: the field is yellow, cereal stems are mainly

Table 3.2 Growers' guide to the dry matter content of whole-crop wheat

DM content of whole-crop wheat (%)	Crop colour	Grain texture
32-35	Green	Soft Brie; some grains milky
36-38	Green	Soft Brie
39-42	Green, ears turning yellow	Soft Cheddar
43-46	Green going yellow	Soft Cheddar
47-54	Yellow, hint of green	Hard Cheddar, with some harder grains
55-65	Yellow, hint of green on stem	Hard Cheddar, with some grains impossible to penetrate with thumbnail
66-70	Yellow/brown, traces of green at nodes	Very hard, with grains impossible to penetrate with thumbnail
71-80	Yellow/brown	Too hard to penetrate with thumbnail; loosening in daytime

yellow, and the grains vary from hard Cheddar to too hard to penetrate with a thumbnail; the growth stages are 87 to 92.

Crop DM 48 to 54%: the overall field colour and the colour of the stems is yellow. Grains vary from very hard Cheddar to too hard to penetrate with a thumbnail, and growth stages are 87 to 92. The DOMD *in vitro* was 50.4% for Aintree and Craig.

Assessing the dry matter content of oats

Oats differ from wheat and triticale in that the oat crop turns yellow and produces hard grains at a lower DM content. Assuming the optimum harvest window is around 50% DM, the crop will be yellow and the grains will vary from very hard Cheddar to too hard to penetrate with a thumbnail. With crops of about 50% DM, the DOMD *in vitro* is much lower for oats than for wheat.

RELATIONSHIP BETWEEN THE ZADOKS GROWTH STAGE, DRY MATTER YIELD AND DRY MATTER CONTENT

Figure 3.1 The relationship between dry matter content and Zadoks growth stage for whole-crop wheat

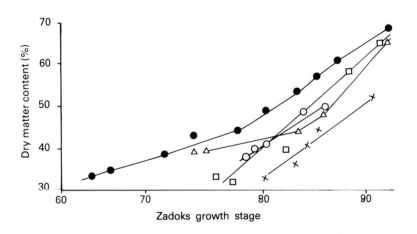

Data from: Wye College ●; Royal Agricultural College ◻; F. Rea △; ADAS ✳; Institute of Grassland and Environmental Research ○

Figure 3.1 shows that the Zadoks growth stage alone was not a useful means of forecasting the DM content. Yields of DM accumulated at widely varying rates in relation to the Zadoks growth stage, depending on local conditions (Figure 3.2).

DRYING RATES OF WHEAT CROPS

Long-term data (Coral *et al*, 1977) from the former Grassland Research Institute (GRI) now the AFRC Institute of Grassland and Environmental Research, suggested an increase in DM content from 20% to 50% between 30 June and 10 August, equivalent to a drying rate of 0.42 percentage units per day. The GRI data also indicated that the drying rate is linear with time.

Figure 3.2 The relationship between dry matter yield and Zadoks growth stage for whole-crop wheat

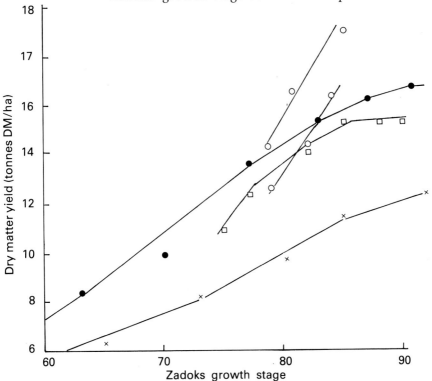

Data from: Wye College ─●─ and ─×─ ; Royal Agricultural College ─□─ ; Institute of Grassland and Environmental Research ─○─

The ADAS data from this 1990 survey showed widely varying drying rates according to site, variety and stage of maturity of the crop, while typical drying rates of 0.45 to 0.75% DM per day were much greater than the values recorded by the GRI.

The sub-samples in each field also varied widely in some cases, particularly close to harvest time; for example, varying from 42.5 to 59.7% DM on the same sampling occasion. This suggests that the crops were not homogeneous, which might cause problems with conservation.

On one farm the crop dried at 2.56 percentage units per day between 18 and 26 July, taking the crop from 43% DM to 63.6% DM. Whilst overall drying rates are likely to be under 0.5% DM per day, rates over 1% and even over 2% DM per day must be expected close to harvest in dry years on dry sites. This has implications for the whole-crop 'harvest window'. It also seems likely that the linear drying rate found at the GRI is not applicable in dry years.

DOMD VALUES *IN VITRO*

Table 3.1 shows DOMD values for wheat and triticale which varied from 63.1% (Lasko at 51.6% DM) to 71.4% (Slejpner at 48.6% DM). Most of the differences were probably due to the natural decline in digestibility with crop maturity and to differences in straw length, but more varietal testing would be of interest. Winter oats (Aintree and Craig) gave DOMD values around 50.4% for crops of about 60% DM, suggesting that oats are inherently less digestible than wheat.

ACKNOWLEDGEMENTS

Thanks are particularly due to the following unpaid farm data contributors: J Alliston, Royal Agricultural College, Cirencester; F Rca, Wotton-under Edge, Gloucestershire; R Weller, IGER, Reading; S Gill, Port Clarence, Middlesborough; G T Atkinson, Winston, Darlington; D Thornley, St Weonards, Hereford; C Wright, Waddesdon, Aylesbury; D Leaver, Wye College, Kent; A Golightly, ICI, Jealotts Hill, Bracknell; A Lister, Liskeard, Cornwall.

Thanks are also due to the sponsors of the survey: The Maize Growers' Association, Mr Gordon Newman, Mole Valley Farmers Ltd, Avon Farmers Ltd, ICI Nutrition Ltd, Dugdales Ltd, and Westward Arable Centres Ltd.

REFERENCES

CORRAL, A. J., HEARD, A. J., FENLON, J. S., TERRY, C.

P. and LEWIS, G. C. (1977) *Whole Crop Forages. Relationship between Stage of Growth, Yield and Forage Quality in Small-grain Cereals and Maize.* GRI Technical Report No. 22, 35pp.

ZADOKS, J. C., CHANG, T. T. and KONZAK, C. F. (1974) A decimal code for the growth stages of cereals. *Weed Research*, **14**, 415-421.

DISCUSSION (Editors' note: this discussion covered both harvesting and ensiling).

At an overall crop DM content of 50%, the straw and ears have rather different DM contents. ADAS hope to develop a sufficiently good visual inspection scheme so that there will be no need to microwave material to assess the DM content.

A problem with filling clamps with whole-crop cereals is that the material is difficult to get into the clamp and it keeps springing up. However, farmers should not feel that they have to keep rolling the clamp until it is consolidated. A Dorset wedge is the best approach. The crop does not need to be chopped short, and clamps which have no effluent control and are thus unsuitable for grass can be used for whole-crop cereals. At 50% DM the density of whole-crop silage is about 3.62 cubic metres per tonne (128 cubic feet per tonne), in clamps of 11 feet deep. There is a difference in density between urea-treated and fermented material due to differences in the DM contents of the crop.

Whole-crop cereals for preservation as fermented material or with alkali should be regarded as different crops. At present fermentation is a better-established technique, but all three current preservation methods – fermented in a clamp, fermented in an Ag-Bag or alkali-treated – can be useful depending on the exact situation. One farmer last year preserved wheat and frost-damaged barley in an Ag-Bag at 48% DM and 50% of the recommended rate of urea, with some harvested after a heavy shower. Results of laboratory analyses, carried out as for grass silage, were surprisingly good and showed a DOMD value of 70%, a metabolisable energy content of 11 MJ per kg DM and a crude protein content higher than for grass silage. There was no smell of ammonia, but the material was very palatable and did not deteriorate in the clamp. It is possible that the crop was very leafy, while the rain could have washed out some grain. Rewetting an over-dry crop can be successful. Some farmers have conserved crops of 75 to 80% DM after heavy rain and achieved a good quality product.

Fertiliser grade urea should not be used for alkali treatment because of the risk of impurities. There is an increasing argument for using fine forms of urea and prills.

CHAPTER 4

WHOLE-CROP CEREALS FOR DAIRY COWS

R H Phipps, R F Weller and J W Siviter
AFRC Institute of Grassland and Environmental Research,
Shinfield, Reading, RG2 9AQ

SUMMARY

First cut perennial ryegrass silage treated with 3 litres per tonne of formic acid was harvested in early June with a dry matter (DM) content of 250g per kg and an in vitro D-value of 660g per kg. Winter wheat was harvested towards the end of June (DM content 500g per kg) for silage and was treated with 4% urea. Maize silage, with no additive, was ensiled at 300g per kg DM in early October. These three silages were fed either as the sole forage to 2-year old Friesian heifers, or as mixtures. The mixtures consisted of either grass and whole-crop wheat, or grass and maize. The whole-crop wheat and maize formed 25, 50 and 75% of the forage ration. In addition, all heifers received 7 kg per day of a concentrate containing 229 g crude protein per kg DM, with a metabolisable energy value of 12.9 MJ per kg DM, during the 20 week experimental period. When compared with grass silage as the sole forage, the incorporation of either whole-crop wheat or maize silage at the 50% inclusion rate in the forage component increased forage DM intake by approximately 1.5 kg per day. However, the incorporation of whole-crop wheat did not increase milk yield. This was in marked contrast to maize, where its inclusion as 50% of the forage increased milk yield by, on average, 1.3 kg per cow per day. Whereas the inclusion of whole-crop wheat tended to increase milk fat content, the use of maize, particularly at the higher inclusion rates, tended to decrease milk fat content when compared with grass silage alone. The inclusion of both whole-crop wheat and maize silage tended to increase milk protein concentration when compared with grass silage alone. Averaged over the three inclusion rates, the highest yield of both fat and protein was obtained by the inclusion of maize silage, the lowest yields of milk constituents were obtained with grass silage as the sole forage, while intermediate values were recorded for whole-crop wheat.

INTRODUCTION

Feed costs are the biggest single input on the dairy farm, and the cost of concentrates is approximately twice that of conserved forage on a dry matter (DM) basis. In a situation where milk output is controlled, most milk producers are attempting to reduce input costs. One such method would be to reduce concentrate input costs and to place greater reliance on cheaper home-grown forage. To achieve this objective, forage quality and intake characteristics must be high. Work at the AFRC Institue of Grassland and Environmental Research (IGER; formerly the AFRC Institute for Grassland and Animal Production) Bernard Weitz Centre, Shinfield has shown that major benefits can be obtained from integrating grass and maize silage.

The aim of the present trial was to provide preliminary data on the effects on forage intake and milk production of incorporating either whole-crop wheat or maize silage into dairy cow rations based on grass silage of high digestibility.

RESULTS AND DISCUSSION

Seventy-nine heifers calved at 2-years old were used in the trial. High digestibility grass silage, maize silage and whole-crop wheat treated with 4% urea were fed as the sole forage or as mixtures. The mixtures consisted of grass silage, with either whole-crop wheat or maize silage forming 25, 50 or 75% of the forage ration. In addition, all heifers received concentrates at a flat rate of 7kg per day during the 20 week experimental period. The concentrate contained 229g crude protein per kg DM and had a metabolisable energy (ME) value of 12.9MJ per kg DM. The chemical composition and nutritive value of the three silages are shown in Table 4.1.

Table 4.1 Chemical composition and nutritive value of grass silage, whole-crop wheat and maize silage

	Grass silage	Whole-crop wheat	Maize silage
Toluene DM (g/kg)	252	515	305
Composition (g/kg DM)			
Total nitrogen	24.8	31.4	12.0
Crude protein	155	196	75
Acid detergent fibre	331	270	292
Neutral detergent fibre	554	456	546
Organic matter			
digestibility *in vitro*	658	709	667
Ammonia-nitrogen			
(% total N)	11.3	47.3	5.9
pH	3.74	7.83	4.02
Gross energy (MJ/kg DM)	18.2	17.7	17.8

Forage dry matter intake

The use of either whole-crop wheat or maize silage to replace part or all of the grass silage led to a substantial increase in forage DM intake (Table 4.2). This occurred in spite of the fact that at the higher inclusion rates (75 and 100%) of whole-crop wheat and maize silage,

Table 4.2 Total forage dry matter intake of dairy cows offered diets containing varying proportions of grass silage and whole-crop wheat or maize silage

% Inclusion of cereal silage	0	25	50	75	100
	Total forage DM intake (kg/day)				
Grass silage with:					
Whole-crop wheat	6.3	7.7	7.6	8.0	7.7
Maize	6.3	7.0	7.7	7.9	7.8

the crude protein content of the whole ration fell below the optimum of 170g per kg DM for early lactation. Thus the increases recorded at the higher inclusion rates may well be an underestimate, and further studies with appropriate protein supplementation are needed.

Milk yield

The incorporation of whole-crop wheat into the ration based on grass silage did not increase milk yield in spite of a marked increase in forage DM intake (Table 4.3). The energy value of the whole-crop wheat may have been lower than that of the grass silage. Results of digestibility studies *in vivo* are awaited, to shed some light in this area, but if this was the case the inclusion of whole-crop wheat would have diluted the energy value of the forage component of the ration. However, because of its higher intake characteristics, total energy intake probably remained relatively constant. It is interesting to speculate that this may be the reason why some farmers consider it important to use molasses with whole-crop wheat.

Table 4.3 Milk yield of dairy cows offered diets containing varying proportions of grass silage and whole-crop wheat or maize silage

% Inclusion of cereal silage	0	25	50	75	100
			Milk yield (kg/day)		
Grass silage with:					
Whole-crop wheat	18.1	18.1	18.2	18.4	18.2
Maize	18.1	18.5	19.4	19.3	18.8

In marked contrast, however, the inclusion of maize silage in the diet led to a substantial increase in milk yield, with a maximum increase of 1.3 kg/day, recorded when the forage component contained 50% maize silage. This result was similar to that recorded the previous year, and should give encouragement to those using 25 to 30% maize silage in their ration seriously to consider increasing its inclusion rate. The decline in milk yield when maize silage was increased from 50 to 100% is attributed to a decline in whole ration crude protein content from 16 to 12%.

Milk composition and yield of milk constituents

Tables 4.4 and 4.5 show the fat and protein contents of the milk and the yields of these milk constituents. The inclusion of whole-crop wheat appeared to increase milk fat content, if it is assumed that the value recorded for 50% grass and 50% whole-crop wheat was anomolous. Although the use of 25% maize in the forage ration increased milk fat content, higher inclusion rates led to a subsequent decline. This was probably due to the increasing grain/starch content of the forage ration and it is interesting to speculate why a similar decline did not occur as the proportion of whole-crop wheat was increased. It is possible that the higher pH of the whole-crop wheat may have led to an improved rumen environment for increased fibre digestion, which could have counter-balanced the additional starch load in the rumen.

The inclusion of either whole-crop wheat or maize silage in rations based on grass silage increased substantially milk protein content.

Table 4.4 Milk composition of dairy cows offered diets containing varying proportions of grass silage and whole-crop wheat or maize silage

% Inclusion of cereal crop	0	25	50	75	100
			Milk fat content (g/kg)		
Grass silage with:					
Whole-crop wheat	41.8	43.4	40.5	42.4	40.7
Maize	41.8	42.4	41.4	40.7	41.1
			Milk protein content (g/kg)		
Grass silage with:					
Whole-crop wheat	32.7	34.3	33.1	33.2	32.5
Maize	32.7	33.9	33.0	33.7	33.1

Table 4.5 Yield of milk constituents of dairy cows offered diets containing varying proportions of grass silage and whole-crop wheat or maize silage

% Inclusion of cereal silage	0	25	50	75	100
	Yield of milk fat (kg/day)				
Grass silage with:					
Whole-crop wheat	0.75	0.79	0.74	0.78	0.73
Maize	0.75	0.76	0.80	0.78	0.77
	Yield of milk protein (kg/day)				
Grass silage with:					
Whole-crop wheat	0.59	0.62	0.60	0.61	0.58
Maize	0.59	0.61	0.64	0.65	0.62

Feeding either whole-crop wheat or maize silage led to increased yields of milk fat. However, the yield of milk protein did not appear to be increased by the use of whole-crop wheat, whereas there was a clear pattern of increasing protein yield as the inclusion rate of maize increased.

CONCLUSIONS

* Both whole-crop wheat and maize silage have high intake characteristics and their integration with grass silage for dairy cow rations will enhance forage intake.

* Some doubt must exist over the energy value of whole-crop wheat, as increased intake was not reflected in increased milk yield. The energy value of the whole-crop wheat would be related to grain loss at harvest, which can be substantial and is almost certainly influenced by the machinery used. Some wheat grain is also lost in the faeces.

* The incorporation of either whole-crop wheat or maize silage into dairy cow rations will improve milk composition. It appears, however, that the main influence of whole-crop wheat is on milk fat, while that of maize silage is on milk protein.

* A general conclusion from the series of trials conducted at the IGER Bernard Weitz Centre is that major benefits can be obtained by replacing a significant proportion of grass silage, normally used in dairy cow rations, by alternative forage sources.

CHAPTER 5

FEEDING CATTLE ON WHOLE-CROP CEREALS

J D Leaver and J Hill

Wye College, University of London, Ashford, Kent, TN25 5AH

SUMMARY

The expansion in the inclusion of whole-crop cereals in diets for dairy cows which took place in the UK in 1989 and 1990 occurred before much research into its feeding value had been carried out. The benefits of whole-crop cereals in terms of high yields, low costs per tonne of production and low storage losses for urea-treated material are clear, but their feeding value is less certain. Research at Wye College into whole-crop wheat harvested at over 50% dry matter (DM) and treated with 4% urea has shown:

* *Digestibility (DOMD) values in vitro of 600 to 650 g/kg;*
* *Considerably greater intakes than for grass silage of similar digestibility for both dairy cows and growing animals;*
* *A depression in milk yield when whole-crop cereals are fed as the sole forage;*
* *No clear effects on milk composition;*
* *An indication that the optimum inclusion rate is 20 to 40% of forage DM intake;*
* *Growth rates in young cattle of 0.50 to 0.75 kg per day when fed alone, and over 1.2 kg per day with 20% concentrates in the diet.*

The poor utilisation of whole-crop cereals when fed alone may be due to the high intake of ammonia from the hydrolysed urea, which has to be metabolised and excreted. The application of urea represents a nitrogen input of about 250 kg per hectare. Studies with whole-crop wheat harvested at an earlier stage, 30 to 40% DM, have indicated substantial reductions in DM yield. However, the digestibility was similar to that of later harvested material (over 50% DM), due to the higher water soluble carbohydrate content compensating for the lower starch level. Current research is examining the use of fermented whole-crop barley silage as a buffer feed in late summer.

INTRODUCTION

Whole-crop cereals are a significant development in cattle production. Whole-crop wheat harvested at a high dry matter (DM) content and preserved with urea has potential attractions as a substitute for grass silage. Whilst grass silage has increased milk yields nationally and allowed greater mechanisation of harvesting and feeding compared with hay, it still has unresolved problems. There are high costs of production, harvesting and storage; environmental problems of effluent and smell; and nutritional limitations to intake and milk production.

Maize silage overcomes many of these problems for farmers in favoured areas, but geographical and climatic limitations are likely to restrict its spread. Whole-crop cereals therefore represent an alternative to maize, but with more widespread application.

Although the conservation of arable crops as silage is not new, the conservation of winter cereals at high yields of DM per hectare is a new concept. The research work carried out at the AFRC Institute of Grassland and Environmental Research, Hurley (See Chapter 1) into the use of urea to permit later harvesting (Deschard, Mason and Tetlow, 1988) has played a significant role in stimulating the current interest in whole-crop cereals. Cutting at a mature stage gives a greater proportion of grain in the total crop, prevents fermentation occurring (reducing storage losses) and produces an alkaline product with potential intake advantages.

Research covering the agronomy, storage, nutritive value and feeding value of wheat and barley has been carried out at Wye College since 1988. The results have demonstrated high DM yields (12 to 16 tonnes DM per hectare), confirmed low storage losses (less than 5% with urea-treated whole-crop cereals harvested at the hard dough stage) and introduced greater flexibility to the farming system. Crops of wheat or barley can be harvested in July if grass or maize silages are likely to be in short supply. The whole-crop cereal can be used as a buffer feed in late summer or as a forage for the winter. Early harvesting also allows either another forage crop to be grown (particularly after winter barley), or additional time for autumn cultivations.

Nevertheless the nutritive and feeding values of whole-crop cereals are less clear due to the lack of research at the present time.

NUTRITIVE VALUE

Species

Most recent research in the UK has been on wheat, together with a small amount on barley. Future research is likely to examine oats, triticale and legumes due to their high yield potential on low input systems. Mixtures of species (especially cereals and legumes) with similar sowing and harvesting dates also seem to have potential. Comparative values for different arable crops are not available.

Varieties

No comprehensive studies have been carried out to compare varieties within species of cereals. There are some indications (Hill and Leaver, 1990) that winter wheat has a higher nutritive value than spring wheat in terms of digestibility and crude protein content.

Height of cut

Stubble height is likely to be an important factor determining nutritive value, due to its influence on the grain : straw ratio of mature crops. The latter can also be influenced by choice of variety or by the use of a straw shortener.

Stage of growth

Until anthesis, cereals show the same changes in nutritive value as grasses. There is an increase in neutral detergent fibre (NDF) and acid detergent fibre (ADF) content, and a decrease in digestibility value and crude protein content. The effect of stage of growth on the nutritive value of whole-crop wheat is shown in Table 5.1.

In this study, as grain fill developed the total DM yield of wheat continued to increase, its starch content increased and its water soluble carbohydrate content declined. However, digestibility remained relatively constant between 30 and 70% DM. Limited studies with winter barley indicated similar changes in nutritive value with time, and a similar digestibility. However, the increase in DM yield between 40 and 60% DM shown in wheat was not apparent with barley, which has implications for the date of harvest and choice of storage method for this crop.

Table 5.1 Stage of maturity and nutritive value of fresh Fortress whole-crop wheat

Date of harvest	13th June	24th June	19th July
DM yield (tonnes DM/ha)	9.6	13.7	15.2
DM (g/kg)	320	450	690
g/kg in DM			
NDF	360	420	490
ADF	230	320	290
Crude protein	110	110	110
Starch	0	110	240
Water soluble carbohydrate	180	60	40
NCD	630	620	630

Source: Hill and Leaver (1991a).

Ensiling or addition of urea

Two systems of storage of whole-crop cereals have developed, either ensiling at 30 to 45% DM content, or storage at 50 to 65% DM using 4% urea to give an alkaline preservation. Ensiled whole-crop wheat showed lower starch and water soluble carbohydrate contents than the fresh material due to fermentation (Tables 5.1 and 5.2); the later the date of cut the higher the starch content and the pH value, but the lower the ammonia-nitrogen (N) content and DM losses in store. Aerobic degradation after opening the silo was also greater in lower DM crops.

Treatment of mature crops with urea (Table 5.2) increased the pH value and ammonia-N content, and reduced DM losses. The digestibility value was similar to that for the fermented material, and urea appeared to have a negligible upgrading effect on digestibility.

Digestibility *in vivo* and *in vitro*

A number of digestibility studies with cattle at Wye College has shown a good relationship between the prediction of digestibility of

Table 5.2 Effect of stage of maturity and of fermentation or urea-treatment on the nutritive value of whole-crop wheat

| Date of harvest | Fermented | | | Urea-treated |
	13th June	24th June	19th July	19th July
g/kg in DM				
Starch	0	90	210	240
Water soluble				
carbohydrate	70	40	10	20
NCD	610	610	620	630
pH	3.9	4.2	6.2	8.9
Ammonia-N (g/kg total N)	90	50	20	390
DM loss in store (%)	13	10	6	2

Source: Hill and Leaver (1991a).

organic matter in the DM (DOMD) from *in vivo* trials and neutral detergent cellulase digestibility (NCD). For both winter wheat and winter barley the DOMD values *in vivo* have been in the range 600 to 650 g per kg.

FEEDING VALUE

A limited number of feeding trials have been carried out in the UK with whole-crop cereals, although substantial research has been carried out in Denmark, particularly with fermented crops (see Chapter 2).

Whole-crop cereals as the sole forage

Initial trials at Wye College examined whole-crop wheat as a forage for dairy cattle and growing cattle.

A crop of spring wheat (variety Axona) was sown in early April 1988, grown on a low input system (cow slurry in winter plus 35 kg N fertiliser per hectare), and harvested in late August at about 65% DM. The estimated DM yield was 11.7 tonnes per hectare. A winter wheat crop (Avalon) was grown in 1989 on a conventional grain production system (including 180 kg N fertiliser per hectare) and, due to the severe drought, harvesting was in late July at about 60% DM. The mean yield was 14.0 tonnes DM per hectare.

In both years the crop was cut with an oilseed rape swather, and picked up with a precision chop forage harvester. The urea was applied in prill form at a rate of 40 kg per tonne DM. Application was on the forage harvester in 1988, and the crop was stored in a conventional unroofed silage clamp sheeted with polythene. The settled density was 276 kg DM per cubic metre. In 1989 the crop was stored in a bottom-unloading tower silo, and the urea was applied at the blower. Studies with mini-silos showed total DM losses of less than 5% when 4% urea was applied to the whole-crop wheat.

The analysis of the two forages when removed from storage is shown in Table 5.3.

Table 5.3 Chemical analysis of whole-crop wheat used in feeding trials in 1988 and 1989

Variety	DM (g/kg)	NDF	ADF	Ash	CP	NH$_3$-N (g/kg total N)	pH	ME (MJ/kg DM)
			(g/kg DM)					
Axona (1988)	628	455	317	63	171	273	8.6	9.6
Avalon (1989)	598	328	210	44	235	289	8.8	11.6

The silages were characterised by their high pH and high ammonia-N contents, arising from the ammonia hydrolysed from urea. The crude protein (CP) content was higher in 1989, probably due to a reduced loss of urea during application. The winter variety (Avalon) had lower NDF and ADF contents than the spring variety (Axona), and consequently had a higher predicted metabolisable energy (ME) content.

A digestibility trial using growing heifers in 1988 gave a DOMD value for the whole-crop wheat of 646 g per kg, which gave a similar predicted ME content to that of 9.6 MJ per kg DM predicted from digestibility values determined *in vitro*.

Feeding trials using mid-lactation cows in a change-over design were carried out with whole-crop cereals in 1988 and 1989. The cows adapted quickly to the forage, in spite of the strong smell of ammonia, but it was noticeable that milk yields declined substantially during the first few weeks after introduction.

The first trial, with Axona, studied firstly the effect of level of concentrate (80% molassed sugar beet pellets and 20% soyabean meal), and secondly the protein content of the concentrate (180 v 240 g/kg crude protein) fed at 6 kg per day. The results for low (6 kg/day) and high (10 kg/day) levels of concentrates are presented in Table 5.4. The higher concentrate level (10 v 6 kg/day) reduced the intake of whole-crop wheat, with a substitution rate of 0.58. The response in milk yield, however, was only 0.5 kg per kg concentrate DM.

Increasing the protein content of the concentrate had no significant effect on the intake of the whole-crop wheat nor on milk yield or composition, but cows increased in liveweight with the 180 g per kg DM crude protein content compared to a small loss in weight with

Table 5.4 Performance of dairy cows fed urea-treated Axona whole-crop wheat with two levels of concentrates

	Concentrate level	
	Low	*High*
Dry matter intake (kg/day)		
Concentrate	5.1	8.7
Whole-crop wheat	14.9	12.8
Milk yield (kg/day)	21.3	23.1
Milk composition (%)	3.92	3.98
Fat	3.17	3.33
Protein		

Table 5.5 Performance of dairy cows fed urea-treated whole-crop wheat

	Axona		Avalon	
Concentrate crude protein (g/kg DM)	*180*	*240*	*165*	*330*
DM intake (kg/day)				
Concentrate	5.2	5.3	3.5	3.5
Whole-crop wheat	14.9	15.0	18.8	19.0
Milk yield (kg/day)	21.3	22.0	17.3	17.2
Fat (%)	3.92	3.95	4.61	4.27
Protein (%)	3.17	3.24	3.68	3.73
Liveweight change (kg/day)	+0.24	-0.05	+0.77	0.69

Sources: Hill and Leaver (1990) and (1991b).

240 g per kg DM crude protein (Table 5.5).

In the second trial, with Avalon, the effect of a higher protein level in the concentrate was examined. The concentrate (fed at 4 kg/day) was based on molassed sugar beet pellets, with 8% soyabean meal (low protein) or 52% soyabean meal (high protein), to give crude protein contents of 165 and 330 g per kg DM respectively. Results are shown in Table 5.5.

The intakes of whole-crop wheat in this trial were again high. The extra soyabean meal in the high protein concentrate had no significant effect on intake or milk yield, but reduced milk fat content. Milk protein content was very high on both treatments.

The most notable feature of these trials with urea-treated whole-crop wheat was the extremely high intakes of the forages relative to their digestibility. However, in both trials the exceptionally high intakes of whole-crop wheat were not reflected in high milk yields or liveweight gain, and a depression of 2 to 4 kg per day in milk yield was noted when the cows were put onto urea-treated whole-crop wheat as the sole forage. Responses to the addition of energy and protein supplements were small.

Also, an examination of the ME intakes (calculated from DM intake and ME content predicted from digestibility *in vivo*), and ME requirements (from recorded animal performance) consistently showed intakes to be higher than requirements. Thus the energy balances indicated 49 MJ ME per day in Trial 2 and 84 MJ ME per day in Trial 1 were unaccounted, suggesting a poor efficiency of conversion of ME, and similar conclusions were reached with youngstock experiments. High blood urea levels (over 60 mg urea per 100ml blood) were found, resulting from the high ammonia intakes, but no adverse clinical symptoms were seen. The discrepencies in energy balance may be associated with the high intake of ammonia, which has to be excreted.

Young cattle fed whole-crop wheat together with concentrates also showed high intakes of forage (2.2 to 2.5% of liveweight) and similar digestibilities of the whole-crop to adult cattle (Table 5.6).

Table 5.6 The intake and digestibility of whole-crop wheat in relation to size of animal

| | *Cattle size* | | |
	Small	*Medium*	*Large*
Initial liveweight (kg)	136	267	340
Dry matter intake			
Concentrate (kg/day)	0.6	0.9	1.1
Whole-crop wheat (kg/day)	4.3	7.4	8.6
Whole-crop wheat (% liveweight)	2.5	2.4	2.2
Whole-crop wheat DOMD (g/kg)	629	623	620

Source: Castejon and Leaver (1991).

Whole-crop cereal mixed with grass silage

Recent studies have examined whole-crop wheat mixed with grass silage. In an experiment where silage replaced up to 44% of the urea-treated whole-crop wheat, milk yield and milk fat content improved with more grass silage in the diet (Table 5.7). Again, in this trial all of the ME intake could not be accounted in animal production, particularly with 100% of the forage as whole-crop wheat.

Table 5.7 Production responses after replacing whole-crop wheat with grass silage in dairy cow diets

	% Whole-crop wheat in forage DM		
	100	78	56
Concentrate DM intake (kg/day)	5.1	5.1	5.1
Forage DM intake (kg/day)	15.1	15.4	15.1
Milk yield (kg/day)	22.3	22.3	23.8
Milk fat (%)	4.15	4.27	4.29
Milk protein (%)	3.31	3.32	3.32
Liveweight gain (kg/day)	0.19	0.13	0.04

Source: Hill and Leaver (1991a).

More recent experiments with 25 to 40% of the total forage DM as whole-crop wheat combined with grass or maize silage have suggested that whole-crop wheat fed at lower levels (3 to 5 kg DM per day) can enhance total intake and milk production, although effects on milk composition were not consistent.

Whole-crop cereal as a buffer feed

The timing of cutting whole-crop cereal in July makes it eminently suitable as a supplement in late summer for cows calving in spring and summer. Research at Wye is currently examining whole-crop barley ensiled at 40% DM as a buffer feed.

CONCLUSIONS

* Whole-crop cereals produce large yields of DM by mid-season at moderate input levels.
* Treatment of whole-crop wheat harvested at 50 to 60% DM has advantages in maximising yield, and producing a stable product with minimal storage losses. There is little evidence that the urea upgrades the feed or that the alkalinity is beneficial.
* The high input of N as urea (equivalent to about 250 kg N per hectare) questions the environmental advantages of this approach.
* For dairy cows the maximum inclusion rate in winter diets

appears to be 3 to 5 kg DM per day.

* Fermented whole-crop cereals have problems of higher DM losses in store, and of aerobic degradation during feeding out. Early cutting also substantially reduces DM yields.

* The nutritive value of fermented whole-crop cereals appears to be similar to that of later cut cereals treated with urea (9.0 to 10.5 MJ ME per kg DM).

* Further research on several nutritional aspects of whole-crop cereals is required to:
   a) study the utilisation of digested energy of whole-crop cereals;
   b) reduce urea usage in late-cut cereals;
   c) determine optimum combinations with other forages in the diet;
   d) assess energy and protein supplements for whole-crop diets.

REFERENCES

CASTEJON, M. and LEAVER, J. D. (1991) *Animal Production*, **52**, 606-707 (Abstract).

DESCHARD, E., MASON, V.C. and TETLOW, R.M. (1988) Treatment of whole-crop cereals with alkali. 4 Voluntary intake and growth in steers given wheat ensiled with sodium hydroxide, urea or ammonia. *Animal Feed Science and Technology*, **19**, 55-66.

HILL, J. and LEAVER, J.D. (1990) *Animal Production*, **50**, 578 (Abstract).

HILL, J. and LEAVER, J.D. (1991a) *Animal Production*, **52**, 606 (Abstract).

HILL, J. and Leaver, J.D. (1991b) Utilisation of whole-crop wheat by dairy cattle. In *Milk and Meat from Forage Crops*, edited by G.E. Pollott. BGS Occasional Symposium No **24**, British Grassland Society.

DISCUSSION: RESEARCH INTO FEEDING CATTLE ON WHOLE-CROP CEREALS

The most important factor with whole-crop cereals found in the Wye trials was the flexibility of the crop. Adding 4% urea had little upgrading effect – there was an increase in DOMD of no more than

10 g per kg. With fermented crops, the earlier the harvest the greater the DM losses and the ensiled material heated more. At any crop DM content, adding urea reduced losses of DM during ensilage and after opening the clamp. Both fermented and urea-treated whole-crop cereals had merits, and there did not seem to be convincing evidence that feeding alkaline, urea-treated whole-crop with an acidic grass silage had advantages in terms of pH.

Whole-crop cereals had been grown at Wye for 3 years. Initial experiments feeding whole-crop wheat led to the conclusion that this was not a suitable forage for feeding alone; its role was in a mixture. When whole-crop wheat was fed with grass silage, there was a better partitioning of energy into milk rather than liveweight gain. With the mixtures, intakes of DM were also very high. Recent studies, not yet reported, had examined the inclusion of lower levels (less than 50%) of whole-crop wheat in the diet, and it appeared that up to 35% or 3 to 5 kg DM per day whole-crop wheat was the optimum inclusion rate in the diet for dairy cows. There was no specific effect on milk composition, but a slight increase in yield. In a trial which had just finished, the substitution of maize by whole-crop wheat led to a slight decrease in milk yield, but milk fat and protein contents improved and liveweight gain increased.

Wye College had not yet looked at the addition of molasses to diets containing whole-crop cereals, but theoretically there might be a good reason for doing so as sugar appeared to 'mop up' excess N in the rumen better than did starch.

In the series of trials reported from Wye, grass silage was not compared to maize silage as, on past evidence, cow performance would have been better with all maize than all-grass silage. Whole-crop wheat did not give as good cow performance as did maize. The cost per unit of DM for urea plus whole-crop wheat was between that of maize and grass.

Wye College results support the digestibility values given in Chapter 7. Digestibility values determined *in vivo* were in good agreement with predictions from NCD.

The college had not specifically investigated if there were advantages in feeding whole-crop wheat to cows other than in terms of milk production and liveweight change.

Diets containing whole-crop wheat do not appear to result in the same increase in milk yield as that observed when maize is fed. In all cases, milk yields were less than predicted, for reasons which are not fully understood. Although there were substantial weight gains for the heifers in the trial at the Bernard Weitz Centre, these were observed for both whole-crop wheat and maize silage, and they did not explain the difference in milk yield response to the two forages.

The proportion of grain in the whole-crop (45 to 50%) was similar to that with maize silage and, presumably, if the digestibility of the material was relatively high there must have been a relatively small loss of grain in the faeces. At Wye College, on average about 12% of the grain fed in the whole-crop wheat appeared as whole grain in the faeces. At the Bernard Weitz Centre few whole grains were visible in the faeces, but on closer examination a considerable proportion of the grains fed had passed through the digestive tract. Whether there are differences in the fermentation of whole-crop cereals and maize in the rumen is unknown, at present, as is the extent to which the energy cost of excreting excess urea could account for the discrepancy between ME intake and output.

There was a variable response in milk protein content to whole-crop silage. At Wye College the forage produced in 1989 was of higher quality than that in 1988 and would be expected to lead to greater production of propionic acid in the rumen, which in turn should increase milk protein synthesis.

During the trial at Wye College when different protein supplements were fed with whole-crop wheat, both soyabean meal and fishmeal were used. In contrast to results expected with grass silage, both milk protein content and milk yield were similar with the two protein sources.

At the Bernard Weitz Centre, the feeding of lucerne rather than grass silage with maize silage has increased DM intakes, but no trials have been carried out feeding lucerne with whole-crop wheat. In practical farming, responses in the yield of milk constituents are more important than increases in DM intake *per se*.

There is little information on differences in feeding value between milling and feed wheat varieties used for whole-crop silages. Growing feed wheat should result in higher crop yields, but there is a need for more information on the optimum stage of maturity to harvest different cereal varieties. It is essential when making large quantities of whole-crop silage to grow several varieties of wheat to allow a realistic harvesting period.

Urea has been applied to the crop at Wye College in 3 ways: through a forage harvester, with a fertiliser spreader at the clamp, and into a tower silo. If distribution is uneven, the ammonia released does not fully percolate the silage and pockets of moulding develop. The question of whether clamps should be rolled has not been resolved.

Urea should not be used on crops with a DM content below 40%. However, propionic acid producing bacteria and enzymes can be used with lower DM crops. The use of stronger alkalis than urea should not be dismissed at this stage in the development of whole-crop cereal silage technology.

There have been no problems of twisted guts at the Bernard Weitz Centre or Wye College with cows fed whole-crop cereals as the sole forage. Animals were fed for up to 12 weeks at Wye College with no clinical problems, although blood urea levels were high. At the Bernard Weitz Centre it has been observed that animals eat mixtures of either whole-crop and grass silage or maize and grass silage more eagerly than grass silage alone.

CHAPTER 6

WHOLE-CROP CEREALS FOR BEEF CATTLE

R M Tetlow [1]
AFRC Institute of Grassland and Environmental Research,
Hurley, Maidenhead, Berks, SL6 5LR
and
J M Wilkinson
Chalcombe Agricultural Resources,
Church Lane, Kingston, Canterbury, Kent CT4 6HX

SUMMARY

The important features of whole-crop cereal silage for beef cattle are:
a) high intake potential compared to wet grass silages;
b) attractive rates of animal growth from relatively low levels of supplementary feed inputs;
c) complementarity with grass silage in mixed forage diets;
d) high crop yields and animal stocking rates.
Wheat is more flexible than barley for conserving as whole-crop silage, with or without treatment with urea, because the grain pericarp is less hard and lignified than is the case with barley. If barley is to be used for beef cattle, it should be harvested no later than the mealy-ripe stage of maturity (30 to 35% crop dry matter(DM)). Urea treatment (4% of crop DM) at harvest confers aerobic stability and can also improve intake, possibly in response to an increase in fibre digestibility in the rumen. The relatively high pH of urea-treated whole-crop silage (pH 8) helps to buffer acidity in the rumen, even in the presence of supplementary grain. Converting both grain and straw into beef through whole-crop silages is an environmentally-attractive way of adding value to cereal crops. At the same time, the relatively high yield per hectare at a single harvest means that margins per hectare from systems based on whole-crop cereal

---

[1]Present address: 1 Woodstock Close, Maidenhead, Berks SL6 7JT

silage are likely to exceed those from grass-based systems of beef production.

INTRODUCTION

Two outstanding nutritional deficiencies require rectifying before whole-crop cereals can achieve their potential as feeds for beef cattle. First, the low protein content of the silages (typically 8 to 10% crude protein in the dry matter (DM)) means that for most feeding situations there is a deficiency of degradable nitrogen in the rumen for optimal digestion by the rumen microflora. Supplementary nitrogen, either as protein or as non-protein nitrogen (NPN), is required.

Secondly, the problem of the instability in air of whole-crop silages must be solved.

In this paper, the factors affecting the nutritive value for beef cattle of whole-crop silages are discussed in the context of the performance of both calves and older beef cattle. Data for crop yield are combined with those for intake and weight gain to indicate the likely levels of output per hectare of land which may be achieved in systems of beef production based on whole-crop silage.

ADDITIVES FOR WHOLE-CROP SILAGES

Ammonia

Ammonia has been used as a source of NPN for whole crop silages. Added at 3 to 4% of the crop DM at harvest, a useful increase of some 4 percentage units in crude protein equivalent may be realised. But early trials indicated that uniformity of distribution of the ammonia is crucial to achieving the required upgrading. Thus Williams, MacDearmid and Innes (1982) found that the performance of beef steers given whole-crop barley harvested at the hard dough stage of maturity and treated with 4% anhydrous ammonia per tonne of crop DM was very poor compared with that from animals given rolled barley and treated straw from the same original crop (Table 6.1). Problems of distributing the gas evenly, and poor palatability were blamed for the disappointing results.

Table 6.1 Performance of beef steers given whole-crop barley
silage made with the addition of anhydrous ammonia
at harvest as the sole feed, or reconstituted rolled
barley and ammonia-treated straw

	Ammonia-treated whole-crop	Rolled barley + ammonia-treated straw
Diet		
Grain:straw	33:66	33:66
Dry matter (%)	40	81
Crude protein (% DM)	21.6	9.56
Intake of DM (kg/day)	3.5	6.2
Liveweight gain (kg/day)	0.12	0.68

Source: Williams *et al* (1982).

Sodium hydroxide

The unpalatability caused by the moulding of whole-crop silages
when exposed to air during the feed-out period may be overcome by
adding a strong alkali, which inhibits mould growth and upgrades
the energy value of the crop at the same time. Sodium hydroxide
(NaOH) will upgrade, but unfortunately it does not confer aerobic
stability (see Chapter 1).

Petchey, Greenhalgh and Mendoza (1980) investigated the nutritive value for beef cattle of mature NaOH-treated whole-crop
barley. Sodium hydroxide was added to the crop at 5% of the DM at
the time of harvest. The treated crop was unstable in air, and at the
advanced stage of maturity (52% DM) digestion of grains was
incomplete. Even so, intake and performance of Friesian steers,
initially 360kg liveweight, given the whole-crop silage supplemented
with soyabean meal was moderate (Table 6.2).

Table 6.2 Performance of beef cattle given mature whole-crop barley treated with sodium hydroxide

	NaOH-treated whole-crop	Rolled barley + NaOH-treated straw
Diet		
Grain:straw:soyabean meal	39:48:13	
Digestibility of DM		
(%) (sheep)	66	64
Intake		
(% of liveweight)	1.95	2.42
Liveweight gain		
(kg/day)	0.72	1.03

Source: Petchey *et al* (1980).

The authors commented that intake of the whole-crop silage was erratic. Growth rate was similar for steers given treated whole-crop and animals given rolled barley, untreated straw and soyabean meal, in similar proportions.

Urea

Urea not only confers stability to the silage (see Chapter 1), it also adds NPN. Further, it is a safe additive which can be applied via the forage harvester. In this way, uniformity of addition can be achieved easily on the farm.

In a recent experiment at Hurley, steers (350kg liveweight) were given mature whole-crop wheat silage (60% DM) made with the addition of different levels of NaOH or urea. It is notable that, even though urea is a weaker alkali than NaOH, intake and digestibility of urea-treated materials were comparable at the same level of additive use, and substantially improved compared to untreated silage (Table 6.3). In this experiment, all animals received a daily supplement of 0.8kg of soyabean meal.

Table 6.3 Nutritive value of whole-crop wheat silages treated with sodium hydroxide or urea

Treatment	Intake of organic matter (% of liveweight)	Digestibility of organic matter in vivo (%)
Untreated	1.7	64
NaOH applied at:		
2% of DM	2.0	65
4% of DM	2.0	67
6% of DM	2.1	68
Urea applied at:		
2% of DM	1.8	66
4% of DM	2.0	69
6% of DM	2.2	69

The greater stability of the urea-treated silages probably gave lower losses during feed-out and this may have contributed to the high levels of intake and digestibility.

Another possibility is that the high pH of urea-treated silage helped to maintain a relatively high pH in the rumen, with beneficial effects on fibre digestion. This feature is discussed in detail in the following sections.

In summary, urea is preferred as the most acceptable additive for whole-crop cereals, since it rectifies the deficiency in nitrogen, and enhances the aerobic stability of the resultant silage.

DIGESTION OF WHOLE-CROP SILAGES

Beef cattle are commonly given supplementary grain, especially during the finishing period. Since whole-crop silages already contain between 40 and 60% grain, it is debatable if additional grain is needed or, if it is, whether it is efficiently used by the animal. Further, the extra grain may depress rumen pH and fibre digestion may be compromised as a result.

In this section, the fate of whole-crop silage in the rumen is considered, with emphasis on maintaining optimal conditions in the rumen for fibre digestion.

Typical compositional values for untreated and urea-treated whole-crop wheat are in Table 6.4. In this experiment, urea was added at 5.66% of the crop DM.

Table 6.4 Composition of untreated and urea-treated whole-crop wheat silages

	Untreated	Urea-treated	SED
Dry matter (%)	59.3	60.3	1.8
pH	4.88	8.91	0.27
Nitrogen (% DM)	0.94	2.79	0.32
NDF[1] (% DM)	50.4	52.8	2.79
ADF[2] (% DM)	29.6	34.1	1.61
Starch (% DM)	30.4	25.7	2.51
WSC[3] (% DM)	3.7	1.2	0.17

[1]Neutral detergent fibre (NDF).
[2]Acid detergent fibre (ADF).
[3]Water-soluble carbohydrate (WSC).

Source: Deschard et al (1988).

Urea treatment was reflected in a large increase in pH and in the content of nitrogen in the silage, whilst differences due to treatment in the contents of structural and non-structural carbohydrates were relatively small. Not surprisingly, urea was reflected in a large elevation in rumen ammonia compared to untreated silage, and the pH of the rumen fluid remained higher throughout the measurement period. Disappearance of neutral detergent fibre (NDF) was faster for urea-treated silage than for untreated material, whilst disappearance of starch was slower for the treated silage.

The consequences of these effects of urea on digestion are shown in Table 6.5. Addition of urea was reflected in a substantial increase in NDF digestibility, but a decrease in starch digestibility. Overall, digestibility of organic matter (OMD) remained unchanged.

Table 6.5 Digestibility of untreated and urea-treated whole-crop wheat silages by beef cattle

	Untreated	Urea-treated	SED
Digestibility (%) of:			
Starch	95.9	88.7	0.52
Cellulose	46.2	66.0	0.32
Organic matter	64.7	66.3	1.79

Source: Deschard *et al* (1988).

PERFORMANCE OF CATTLE GIVEN UREA-TREATED WHOLE-CROP SILAGES

Summarised in Table 6.6 are feed intake and liveweight gain data from recent experiments in which urea-treated whole-crop silages were given to beef cattle of different initial age and weight. In all the experiments, soyabean meal was included in the diet at 0.4% of liveweight (calves) or 0.2% of liveweight (older cattle).

Intake of wheat silage by calves was exceptionally high, and in one experiment greater than that of triticale by comparable animals. In all cases there was a response in liveweight gain to the inclusion of supplementary rolled barley in the diet.

Mixtures of grass silage and untreated whole-crop wheat silage were examined in one experiment with 300kg Friesian steers. The results for intake, digestibility of DM and liveweight gain are summarised in Table 6.7.

It is notable that the grass silage used in this trial was of very high digestibility, and was capable of supporting a liveweight gain of 1 kg per day with supplementary soyabean meal at 0.8kg per head per day (0.4% of liveweight).

Dry matter intake showed little change as the proportion of grass silage in the diet was increased, but the weight gains achieved by the calves increased markedly in response to the increase in digestibility of the diet.

The optimal mix of urea-treated wheat silage and grass silage would appear to be around 75:25 grass:wheat, depending on the

Table 6.6 Intakes and liveweight gains of beef cattle given urea-treated whole-crop cereal silages

	Rolled barley (% of liveweight)	DM intake (% of liveweight)	Liveweight gain (kg/day)
Wheat			
Calves 120kg liveweight			
Tsatsu (unpublished)	zero	2.95	0.83
Calves 150 to 220kg liveweight			
Tetlow (unpublished)	zero	2.70	0.76
	0.4	2.76	1.10
	0.8	2.86	1.02
Steers 300 to 400 kg liveweight			
Deschard *et al* (1988)	zero	1.83	0.81
	0.6	2.20	1.03
Tetlow (unpublished)	zero	1.87	1.05
	zero	1.85	0.81
	0.4	2.24	1.03
Triticale			
Calves 150 to 220kg liveweight			
Tetlow (unpublished)	zero	2.03	0.70
	0.4	2.35	0.82
	0.8	2.39	1.03

digestibility of the grass silage.

It remains to be seen whether or not the inclusion of soyabean meal is critical to the achievement of the performance levels recorded in the above experiments. It is more likely to be needed in diets for beef calves less than 300 kg liveweight, than for older animals.

BEEF SYSTEMS

The performance levels achieved in the above experiments suggest that acceptable levels of growth in beef cattle should be achievable

Table 6.7 Performance of beef steers given mixtures of grass silage and urea-treated whole-crop wheat silage

Proportion of grass silage in diet[1]	DM intake (% of liveweight)	Digestibility of DM (%)	Liveweight gain (kg/day)
0	2.41	64	0.68
25	2.46	66	0.83
50	2.46	72	0.90
75	2.50	76	1.00
100	2.44	78	1.00

[1] Soyabean meal was given at 0.8kg per head per day.

from urea-treated whole-crop cereal silage in lifetime systems of production, or in store cattle finishing, with relatively low levels of supplementary feeding. Coupled with the high yields of wheat and triticale, potential outputs per hectare (Table 6.8) look attractive, and should be reflected in acceptable margins, compared with other intensive systems of beef production.

The number of animal days per hectare was highest for the zero-grass diet, reflecting the higher DM yield of the whole-crop wheat. Output of liveweight gain per hectare was highest with 75% grass silage in the mixture, reflecting the superior daily liveweight gains from the grass silage compared to the wheat silage. Liveweight gains per hectare were similar for the diet comprising 25% grass silage and that containing 50% grass silage in the mixture.

Inevitably, comparisons such as these are of limited value, since they only relate to the respective qualities of the two types of silages used in the experiment described in Table 6.7, and to the assumed yields in Table 6.8. But the data indicate that there may be flexibility in the proportion of grass to whole-crop silage in the diet of growing beef cattle, with relatively little change in output per hectare. Thus the proportion of whole-crop silage on the farm may be determined by other factors, such as suitability of fields, or desired crop rotations.

The cost of growing whole-crop silage is likely to be similar to that

Table 6.8 Estimated output per hectare by beef cattle given mixtures of grass silage and urea-treated whole-crop wheat silage[1]

	Grass silage (% of total silage DM)				
	0	25	50	75	100
Assumed yield of silages (tonnes DM/ hectare)	14	13	12	11	10
Animal days/hectare	*1647*	1448	1274	1127	1050
Liveweight gain/hectare (kg)	1208	1296	1318	*1382*	1288

[1] Feed intake and liveweight gains as in Table 6.7.

of cereal grain, since the inputs to grow the crops are likely to be the same. The choice of which area to cut for silage will probably be made on the basis of prospective grain yield and quality.

However, compared to grass silage, annual harvesting costs will be approximately one third those of a three-cut grass silage system, since only one harvest occurs per annum with whole-crop cereals.

The high DM content of whole-crop cereal silages means that no effluent is produced, and therefore there is no need for effluent collection, storage or disposal.

CONCLUSIONS

In conclusion, whole-crop cereal silage offers the beef farmer opportunities for achieving high levels of output of liveweight gain per hectare from systems of production which are likely to require lower levels of working capital to produce the forage feed than is the case with conventional grass silage. In a situation of high interest rates, this advantage may be crucial to the maintenance of acceptable levels of profit from intensive beef production.

REFERENCES

DESCHARD, G., MASON, V.C. and TETLOW, R.M. (1988) Treatment of whole-crop cereals with alkali. 4. Voluntary intake and growth in steers given wheat ensiled with sodium hydroxide, urea or ammonia. *Animal Feed Science and Technology*, **19**, 55-66.

PETCHEY, A.M., GREENHALGH, J.F.D., and MENDOZA, R.F. (1980) Alkali treatment of whole-crop barley. *Animal Production*, **30**, 489 (Abstract).

WILLIAMS, P.E.V., MACDEARMID, A., and INNES, G.M. (1982) Anhydrous ammonia-treated whole-crop barley as a diet for beef steers. *Animal Production*, **34**, 381 (Abstract).

DISCUSSION

In the trials reported, soyabean meal was given to the animals receiving urea-treated silages because these silages were being compared to NaOH-treated material, which required additional rumen-degradable protein. The need for a protein supplement when beef animals are fed on urea-treated whole-crop silages has not been studied, and it is likely that the nitrogen from urea would be adequate, without soyabean meal, for beef cattle of more than 300kg liveweight.

It was disappointing that when beef cattle were given a combination of whole-crop and grass silages, intake did not increase markedly compared to grass silage alone. This may, however, have reflected the very high quality of the grass silage.

If whole-crop silage is fed rapidly and comprises a high proportion of the diet, either fermented or urea-treated silages may result in similar levels of animal performance. There may be an advantage in urea treatment due to the alkalinity of the resulting material and its aerobic stability in situations, such as with dairy cattle, when improvements in milk quality are desired and when whole-crop silage forms a lower proportion of the diet.

At Kites Nest Farm (see Chapter 9) the aerobic stability of urea-treated silage allows the forage to be given to the beef cattle only twice a week.

The performance of beef cattle given fermented whole-crop barley made in Ag-Bag silos was compared to that of barley-beef animals in a farm trial in 1982. Daily liveweight gains were similar for both groups, but the area of land required for the animals given the whole-crop silage was only half of that required for cattle on the barley grain diet.

When the performance of beef cattle given whole-crop wheat silage treated with NaOH or urea was compared (Table 6.3), the trial was a Latin Square design, and consequently daily liveweight gains could not be measured.

CHAPTER 7

NUTRITIONAL EVALUATION OF WHOLE-CROP WHEAT

A H Adamson
ADAS Nutrition Chemistry Department,
Burghill Road, Bristol, BS10 6NJ
and
A Reeve
ICI Nutrition, Alexander House,
Runcorn, Cheshire, WA7 2UP

SUMMARY

With whole-crop cereals variations in cutting date, cutting height, grain losses, method of ensilage and efficiency of conservation, superimposed on variations in the quality of the standing crop which is harvested, not surprisingly result in the production of feeds of widely different nutritional values. The analytical results presented to the producer may also include a sampling error for such a heterogenous material. A simple calculation of feed metabolisable energy (ME) values for straw:grain ratios of 50:50 or 60:40 shows a difference of around 0.8 MJ ME per kg dry matter (DM). Data have been collated for 67 samples of whole-crop cereals which have been analysed by ADAS during normal advisory work throughout England; the crops were ensiled by conventional fermentation (FWCC) or with urea (UTWCC). These data show a wide variation in the DM contents of both FWCC and UTWCC, from 230 g per kg DM (FWCC sample) to 812 g per kg DM (UTWCC sample), and also a wide range in the crude protein and ammonia contents of UTWCC, which is of considerable concern, in particular for crop preservation quality, efficiency of ME utilisation and excretion of nitrogen. The relatively narrow pH range of FWCC reflects the ease of obtaining a satisfactory fermentation with this type of material; the generally high but wide range of pH values for UTWCC is in agreement with the ammonia values. Both types of material had a very wide range of cell wall contents, while the straw content of the UTWCC was higher and more variable than that of FWCC. The ME value was predicted from routine laboratory analysis using an equation developed for maize silage, in the absence of a specifically derived relationship for whole-crop material. In general, the ME content of the FWCC was about 1 MJ per kg DM lower than that of UTWCC. The ME values were generally low and in agreement with data obtained elsewhere.

INTRODUCTION

There are several factors which affect the chemical composition and nutritive value of harvested whole-crop cereals, whether these be preserved by conventional silage fermentation techniques or by the addition of preservatives which inhibit fermentation. The most common preservative used in England at the present time is urea, the aim of which is to convert it to ammonia by naturally occurring urease enzymes. The level of urea application recommended is of the order of 40 to 50 kg per tonne of dry matter (DM).

The proportion of grain to straw is the major source of variation in the chemical composition of harvested whole-crop cereals, due to inherent variations in the length of straw, variations in the yield of grain, variable cutting heights and grain losses at harvesting. Table 7.1 shows how the calculated metabolisable energy (ME) values vary with different grain:straw ratios, based on average energy values, and ignoring the variation in the ME value of the grain or straw fractions. These calculations suggest that preserved whole-crop wheat will seldom contain a high concentration of nutrients.

Table 7.1 Effect of grain:straw ratio on the calculated metabolisable energy value of whole-crop wheat

Grain:straw ratio	100:0	60:40	50:50	40:60	0:100
Mean ME (MJ/kg DM)	13.7	10.7	9.9	9.1	6.1

Since the lack of consistency in the feeding value of whole-crop cereals is not always recognised, it was decided to collate analytical data from commercial farms for whole-crop wheat and to supplement this with available measurements determined *in vivo*, particularly ME value.

Data were collated for a total of 67 samples which had been analysed in the normal course of advisory work throughout England. Virtually all samples were analysed for DM, crude protein, ammonia content, digestibility by neutral detergent cellulase (NCD), and pH value, with variable numbers of samples being analysed for neutral detergent fibre (NDF) and starch content. The methods of analysis were those used routinely in ADAS laboratories.

RESULTS OF THE FEED ANALYSES

Data were obtained for the composition of whole-crop wheat ensiled by a conventional fermentation process (FWCC) or treated with urea (UTWCC). No attempt was made to distinguish between FWCC samples which had been treated with an additive or left untreated, or the level of urea applied to samples of UTWCC.

The mean chemical composition, standard deviation and range of values are shown in Table 7.2.

There was very wide variation for all parameters of UTWCC, but a somewhat narrower range of crude protein, ammonia and pH values for FWCC, which was a generally well-fermented product.

Table 7.2 The mean chemical composition of whole-crop wheat silages

	DM (g/kg)	Crude protein (g/kg DM)	Ammonia-N (g/kg TN)	pH	NDF (g/kg DM)	Starch (g/kg DM)	ME (MJ/kg DM)
Urea-treated whole-crop wheat							
Mean	549	244	250	7.7	438	224	10.3
SD ±	97.8	68.7	82.4	1.03	66.2	77.1	0.82
Range	376 to 810	120 to 513	90 to 448	4.7 to 8.8	249 to 577	91 to 386	8.7 to 12.2
Fermented whole-crop wheat							
Mean	337	102	70	4.0	483	109	9.3
SD ±	78.8	20.7	25	0.31	148.7	43.8	0.35
Range	230 to 512	71 to 147	40 to 130	3.5 to 4.9	323 to 617	78 to 140	8.4 to 9.8

The FWCC had a higher NDF content and lower DM and starch levels; calculated ME values were lower than for the UTWCC. Protein levels, ammonia content and pH levels were also much lower in FWCC than in UTWCC.

The ME content was calculated from the digestibility value determined by neutral detergent cellulase (NCD) using a relationship developed for maize silage because of the absence of a specific relationship between any laboratory measurements and ME values for whole-crop cereals. The ME and starch contents did not show a close relationship, although starch levels accounted for 45% of the variance in ME.

CHEMICAL COMPOSITION AND NUTRITIVE VALUE

Dry matter

Figure 7.1 shows a wide range of DM contents for both UTWCC and FWCC. One value was as low as 230 g per kg for FWCC, which is a level at which some effluent could be produced, and this demonstrates the caution which must be exercised by those who ignore possible effluent production from whole-crop cereals. At the maximum DM level of 812 g per kg, close control would be required to

Figure 7.1 The dry matter content of fermented and urea-treated whole-crop cereal samples

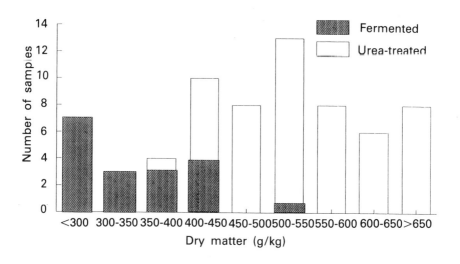

prevent overheating when anaerobic fermentation was the intended method of conservation.

The average DM content of UTWCC was within the recommended target level of 450 to 600 g per kg but the extreme levels were outside this range. Tetlow has indicated (Chapter 1) that the fermentation of wetter crops, coupled with the addition of alkali, buffers changes in pH and prevents an adequate level of acidity being reached for satisfactory conservation, so that a clostridial fermentation may result. At the other extreme, the urease activity in very dry material (in excess of about 700 g per kg DM) is likely to be limited and, indeed, the ammonia- nitrogen (N) per kg of total nitrogen (TN) in UTWCC samples with DM levels of 810, 783, 736 and 681 g per kg were low, at 100, 70, 130 and 100 g per kg respectively.

Crude protein and ammonia-nitrogen

The inherently low crude protein content of cereals and straw is reflected in the relatively low values for FWCC and the relatively narrow range. The proportion of crude protein catabolised to ammonia was also relatively small, demonstrating a consistently acceptable fermentation on the accepted basis that 100g ammonia-N per kg TN indicates a well-fermented product.

Figure 7.2 The crude protein content of samples of urea-treated whole-crop cereals

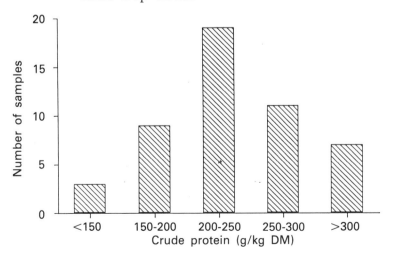

Figure 7.3 The ammonia-nitrogen content of samples of urea-treated whole-crop cereals

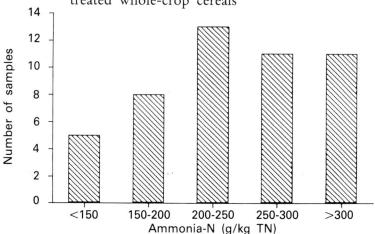

The distributions of crude protein and ammonia contents in UTWCC are shown in Figures 7.2 and 7.3 respectively.

The addition of nitrogen in the form of urea clearly increased the crude protein content, and the conversion of urea into ammonia by naturally occurring urease enzymes was reflected in the high ammonia content. The very wide range in ammonia content is of concern and there are a number of possible explanations:

* A wide variation in the overall levels of urea application.
* Uneven urea application and sampling from pockets of low or high concentration.
* Low urease activity due to high DM contents, as noted previously.

Clearly these variable levels have practical implications for UTWCC which relies on ammonia for preservation and increased shelf life. The low levels may well be ineffective under aerobic conditions during feed-out, and hence result in increased spoilage. Whilst high levels of ammonia are not likely to be toxic in clinical terms, there is an appreciable energy cost to excreting the excess ammonia which cannot be utilised by the animal. Also, the increased level of N in the resultant slurry needs to be recognised for efficient utilisation as a fertiliser.

pH

The relatively narrow range of pH values for FWCC is a reflection of the ease of obtaining a satisfactory fermentation with this type of material. The generally high but wide range of pH values of UTWCC, as shown in Figure 7.4, is in agreement with the generally wide range of ammonia contents. Eighty-five percent of samples of UTWCC had pH values above 7.0, indicating a generally alkaline material.

Figure 7.4 The pH of samples of urea-treated whole-crop cereals

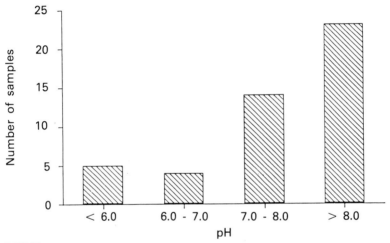

NDF

The probable variation in grain/straw ratio and date of cutting is reflected in the very wide range of cell wall contents, as measured by NDF values, both for UTWCC and FWCC. Figure 7.5 shows that the majority of samples lay between 249 and 450 g per kg DM. The close relationship ($r^2 = 0.77$) between starch and NDF contents, which reflects the proportion of grain to straw, is not surprising. This has important implications for ration formulation which optimises starch and/or NDF content, particularly where average rather than actual values are used.

Starch

The earlier-cut FWCC showed low starch values and a narrow

Figure 7.5 The neutral detergent fibre (NDF) content of samples of urea-treated whole-crop cereals

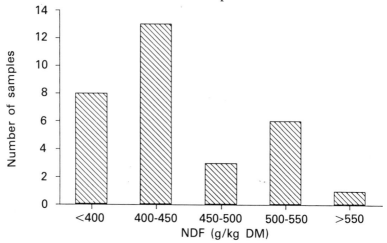

range compared to the more mature UTWCC (Figure 7.6), where the wide range indicated a much greater variation in the contribution from grain. This also has important implications for the efficient utilisation of nutrients, although the rate of rumen degradation of starch fed mainly as whole grain is unclear. Whilst the average starch content agrees with that given in Chapter 1, there were some very high values and the highest level of 386 g per kg DM was related to

Figure 7.6 The starch content of samples of urea-treated whole-crop cereals

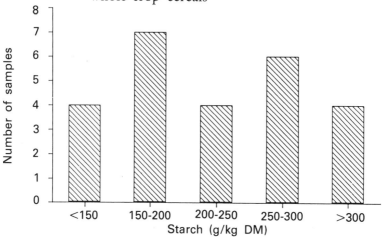

an NDF value of 249 g per kg DM, a digestibility *in vitro* of 700 g per kg DM and a calculated ME value of 12.2 MJ per kg DM. This would approximate to a grain content of 80%.

Metabolisable energy

Since there are no satisfactory means for calculating the ME value of preserved whole-crop cereal from routine laboratory analysis with an acceptable degree of accuracy, the equation developed by Givens (1990) in part association with the Maize Growers' Association was applied to all whole-crop wheat. This equation predicts the digestibility of organic matter in the DM (DOMD) from digestibility in neutral detergent and cellulase (NCD) and calculates ME on the basis of an assumed ratio of DOMD:ME of 0.157.

The FWCC samples (Figure 7.7) showed a low mean ME value of 9.3 MJ per kg DM, with a maximum value of 9.8 MJ per kg DM, whilst the UTWCC samples (Figure 7.8) showed a mean value 1 MJ higher but an even distribution over a wide range. The maximum value of 12.2 MJ ME per kg DM for one sample of UTWCC corresponded to very high starch and low NDF contents, suggesting a high proportion of grain. These figures agree with those obtained by Kristensen (Chapter 2) and Leaver and Hill (Chapter 5).

Figure 7.7 The metabolisable energy (ME) content of samples of fermented whole-crop cereals

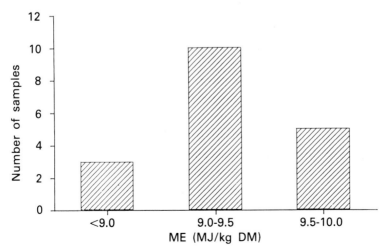

Figure 7.8   The metabolisable energy (ME) content of samples of urea-treated whole-crop cereals

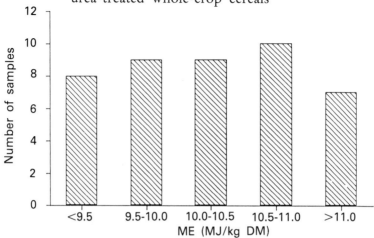

It is an accepted fact that the energy value of the digestible organic matter (DOM) of fermented feed is higher than that of unfermented products (Givens *et al*, 1989). UTWCC is not a fermented product and the DOMD:ME ratio would be expected to be less than the value of 0.157 applied to maize silage. Indeed, the measured DOMD and ME values from one sample examined at the ADAS Feed Evaluation Unit have indicated a ratio of 0.15. This suggests that the ME values calculated according to the equation applied to maize silage may be about 5%, or 0.5 MJ per kg DM, above the true values.

CONCLUSIONS

Data for conserved whole-crop wheat collated from advisory work on commercial farms indicate that the product is very variable, particularly urea-treated material. This must be recognised when formulating rations including whole-crop cereals, since variation occurs not only in the total N fraction but also in the proportion of N present as ammonia, and there are variations in the energy content, whether this is derived from starch or digestible fibre.

On average, whole-crop wheat is not a forage of high nutritive value and the indications are that the fermented product has a low ME content, although the limited range of the present study is recognised. Variations in starch levels strongly suggest that the major cause lies in the different proportions of grain to straw.

There are no indications of any particular problems in achieving a successful fermentation or preservation. However, the N data indicate a need for closer control of urea application, whilst the range in DM content illustrates the need for a more precise estimate of moisture content at harvesting.

The advantages of whole-crop cereals are likely to lie in agronomic factors such as total forage production and farming systems to maximise forage production, rather than any special nutritional properties.

REFERENCES

GIVENS, D.I., EVERINGTON, J.M. AND ADAMSON, A.H. (1989) The digestibility and metabolisable energy content of grass silage and their prediction from laboratory measurements. *Animal Feed Science and Technology*, **24**, 27-43.

GIVENS, D.I. (1990) Proceedings of Maize Growers' Association Conference, Berkshire College.

DISCUSSION

The Danish advisory service uses a conversion factor of 0.15 to predict ME from DOMD, and in general the information from Denmark given in Chapter 2 confirms UK data. Even for samples with similar grain:straw ratios and starch contents, ME values may vary considerably as there can be marked differences in the ME content of the straw fraction.

The conversion factor used by ADAS for predicting the ME content of maize from its digestibility value is 0.157. However, within ADAS, a factor of less than 0.15 is used to predict the ME content of straw, while values greater than 0.15 are used for feeds which undergo much fermentation, and at the first whole-crop conference at Hurley in 1990, discussion indicated that the most appropriate factor was around 0.15 for fermented whole-crop silages.

Starch content comprises a very important part of the analysis of whole-crop material, but this determination is often not carried out routinely. Most of the starch is degraded relatively slowly in the rumen. The starch content of whole-crop cereals in Denmark is about 25 to 27%, which is much higher than values for the UK. A starch content of 25 to 30% would be preferred for late-cut, whole-crop material in the UK.

The generally high intakes of whole-crop cereals are probably more important nutritionally than the energy content of the material, which is moderate. However, if a forage of high intake potential, such as maize silage, is already being fed the introduction of whole-crop cereals is unlikely to increase further DM intake. The likely effects of feeding whole-crop cereals must be assessed after studying the overall diet composition.

The urea-treated whole-crop may not have such a high buffering capacity as sometimes thought. The beneficial effects on pH may be mainly because an alkaline not an acid product is fed. The ammonia-N content is not an indication of the fermentation quality of urea-treated material.

CHAPTER 8

MACHINERY FOR WHOLE-CROP CEREALS

P L Redman
ADAS Mechanisation Unit, Wrest Park, Silsoe,
Bedford, MK45 4HS
and
A C Knight
AFRC Institute of Engineering Research,
Wrest Park, Silsoe, Bedford, MK45 4HS

SUMMARY

Commercially, harvesting whole-crop cereals for silage is at an early stage, and information on the performance of harvesting equipment is very limited, being available only from development work and pioneer users. There are many similarities, but also some key differences, compared to grass silage. As whole-crop cereal silage is likely to be produced on farms which already conserve grass silage, economic considerations suggest that, at least at present, whole-crop harvesting systems should be based on existing equipment used for grass silage. The standard chopping mechanisms of forage harvesters appear to be capable of effectively chopping whole-crop material to the target chop length of 56mm. There are two basic options for achieving the required swath working width for whole-crop material, namely harvesting direct with a wide header, or picking up pre-mown material. The effect of the chosen system on grain losses must be considered, as these may be substantial. The lower bulk density of whole-crop cereals has a marked effect on the capacity of the transport system which may, if set up for grass silage, limit the overall system output. Precautions must be taken to avoid significant losses of material both when filling trailers in the field and at all handling stages up to feeding. Development work on equipment to harvest whole-crop cereals is currently in progress, including studies on a novel pick-up for attachment to forage wagons, big balers and forage harvesters. An adaptation of this pick-up which includes a reciprocating cutterbar will facilitate direct harvesting of the crop.

Research is also needed to develop methods of applying urea in the dry form at relatively high rates.

INTRODUCTION

As the technique of harvesting whole-crop cereals is in the early stages of commercial adoption, the available data on performance and guidance on design and operation can only be based on experience obtained during development studies and from pioneer users. Whole-crop cereals produce a bulky fibrous material, and there are many similarities with the harvesting, handling and storage of grass silage. Furthermore, as this crop will be produced on farms already geared up for grass silage, it is economically imperative to take existing silage equipment as the basis for a whole-crop harvesting system. In the short term at least, the demand for specialised equipment, even if required, is unlikely to tempt manufacturers into a full scale development programme.

As far as machinery is concerned, whole-crop cereal silage differs from grass silage in the following key respects and these must be taken into account when devising a system for harvesting, handling and storage based on grass silage equipment.

(i)   No wilting stage is required, so that harvesting can take place direct from the standing crop. Typical harvested yields of 20 to 25 tonnes per hectare (Tetlow, private communication) are broadly equivalent to those of grass silage.

(ii)   Due to their different flow characteristics, there is a tendency for the grain and straw fractions to separate at each handling stage and there is a risk of significant grain losses, particularly during harvesting.

(iii)   The fibrous fraction does not readily compact prior to the softening action of alkali treatment. Consequently, the bulk density of whole-crop material (90 to 150 kg per cubic metre) (Boyd and Longmuir, 1983) is only half that of grass silage (typically 270 kg per cubic metre) (ADAS, 1983). Although this is disadvantageous for transport and storage, the open structure allows percolation of ammonia. Due to differences in dry matter (DM) content, bulk density expressed on a DM basis is comparable with that of grass silage (70 kg DM per cubic metre).

(iv) Treating whole-crop cereals with urea at the rate of 4 to 5% involves the handling of appreciably larger quantities of additive compared to conventional granular silage additives (>5 times). This requires a fresh approach to application and distribution although, due to the mobility of the ammonia produced, there may be some tolerance in terms of uniformity of distribution.

(v) There is virtually no risk of effluent production and the treated material is biologically stable, so that it is not necessary to maintain anaerobic conditions.

The following sections identify the practical consequences of these differences and indicate some of the compensatory measures which can be taken, drawing on evidence from grass silage systems, limited specific data, and the experience of those developing systems in practice.

HARVESTING

Various reports over the last decade or so indicate that the standard chopping mechanisms of forage harvesters will effectively chop straw or whole-crop material to the target chop length of 56 mm.

For a given chop length setting and machine condition, forage harvester output is determined by a combination of the weight of crop presented to the harvester and its forward speed, within the limits of power available. Extensive survey data of grass harvesting systems (ADAS, 1983) indicate that forage harvesters are rarely driven faster than 2 metres per second (4.5 mph). This is most probably due to the risk of blockage from irregular swaths and the need to operate within the control limits of the machine combination. There is every reason to suggest that similar restrictions will apply to cereals, even when harvesting a standing crop. Wider working widths, the risks of fouling the cutter/pick-up and the difficulty of controlling trailed configurations are all likely to limit forward speed.

Consequently, it is the quantity of crop presented to the harvester which will determine work rate, and for a given crop yield effective working width becomes the dominating factor. In practice, modern trailed forage harvesting systems can perform at spot throughputs in the range 30 to 35 tonnes per hour (ADAS, 1983) with grass, although only 60 to 70% of this output (say 20 tonnes per hour) will be consistantly achieved due to the inevitable delays of changing trailers, etc. Similar outputs are reported with whole-crop cereals. To sustain this output with a crop of, say, 25 tonnes per hectare within the limit of 2 metres per second forward speed, an effective working width of 2.4 metre is required, increasing *pro rata* for lower yielding crops. Although for crop quality there may be less pressure to harvest within a short period, the effective use of machinery and labour is always desirable, and the risk of grain loss increases as the crop matures.

The two basic options for achieving the required working width are:

i)   Harvest direct with a wide header;
ii) Pick up pre-mown material.

Direct cut header attachments are available for most 'metered' chop harvesters, at the required width. One study (Boyd and Longmuir, 1983) reported problems with blockages and observed appreciable grain loss due to the aggressive and rotating action of the discs. Significant quantities of grain were also lost from gaps in the housing of the chopping and feed mechanism until blanking plates were fitted.

For these reasons, the more gentle action of a cutterbar is preferred, in conjunction with a conventional header from a combine harvester which has been designed for this situation and which will cope adequately with a standing crop. These are reported to be working effectively in practice. The width of centre feed type header which can be fitted to a trailed harvester is, however, limited by the need to accommodate the drawbar. No data on the loss of grain by harvesting with this method can be identified. It is, however, worth noting that typical grain losses for conventionally harvested wheat grain crops are 150 kg per hectare (ADAS, 1982); grain at the conventional harvesting stage is, of course, much more susceptible to this type of loss.

The alternative approach is to pre-mow the crop prior to picking up with a standard forage harvester configuration. There is clearly a risk of losing grain both during cutting and picking up. Early studies indicated that this could be of the order of 6 to 10% of grain DM yield (Wilkinson, 1981). Mowers 2.4 metre wide are commonly available and, with high yielding crops, will produce swaths of sufficient density. Light crops may need swaths to be combined from more than one pass, using the gentle swathing conveyors attached to some models of mower, trapping grain in the process, although the aggressive rotary action is still likely to cause some losses.

Self-propelled harvesters commonly used by contractors have many attributes for this purpose. They have higher throughputs (+75%) which can be supplied by wider headers at higher forward speeds.

In theory, roll balers have the advantages of compacting the crop and of being a multi-stage operation, but one preliminary observation indicated that the friction from the rolling mechanism induces very heavy grain losses (30%) during bale formation.

TRANSPORT

The lower bulk density of whole-crop cereal material, as harvested, is approximately half that of grass, which will have a marked effect on the capacity of the transport system. If this is sized for grass silage, then transport is likely to limit the overall system output when working on whole-crop material.

Harvesting systems designed for industrial scale operation incorporate a load compression facility. This is unlikely to be justified for farm-scale operations.

There are various reports of significant losses, both of light crop fractions and of grains, during the filling of trailers. The aim is to allow entrained air to escape whilst trapping these fractions. Attention should be paid to ensuring that the rear gate is grain-tight. It has been suggested (Boyd and Longmuir, 1983) that the distance between forage harvester spout and trailer should be kept to the minimum to avoid wind losses by towing trailers alongside, and at the same time increasing the height of the mesh canopy. Although consolidation to higher bulk densities will improve the efficiency of transport and storage, it is also claimed that a more open structure facilitates the percolation of ammonia. The ideal chop length therefore remains a matter for debate.

APPLICATION OF UREA

The relatively high application rate of urea (5%) requires careful planning. In addition to the logistical problems in supplying a chemical at rates which may exceed 1 tonne per hour, current proprietary silage additive applicators have a range well below the capacity required. Developments for the application of absorbent additives will be more suitable for applying urea.

Application in the prilled form minimises the weight to be handled but the urea is susceptible to some separation during handling, before retention by the moist crop. Ideally, handling should be based on some form of intermediate bulk container such as 0.5 tonne big bags. It is not feasible to construct handling hoppers of sufficient capacity on the forage harvester to allow delivery by gravity. A more logical approach is to mount the hopper so that it has a lower centre of gravity and to use a metered feed pneumatic conveyer to transfer the prills to the crop flow in the forage harvester. Applying urea in the form of a solution adds to the quantity to be handled, but allows pumping techniques to be used for transfer and application. As proprietary systems have inadequate capacity, these will have to be assembled from separate components.

GRAIN SEPARATION

Grains and chopped straw have different flow, aerodynamic and dimensional properties – characteristics which are exploited to bring about separation in combine harvesters.

All handling stages up to feeding should be devised and operated to limit these effects. Pneumatic conveying to conical heaps at the natural angle of repose is particularly conducive to separation. Tractor loading systems skilfully used can keep separation to the minimum by keeping shallower angles of repose, where grain is not allowed to roll. Tractor loaders can be used with a scooping action so as to recombine any separation at each handling stage.

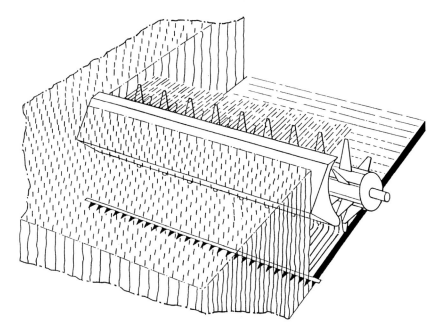

Figure 8.1 Schematic diagram of the high speed pick-up with reciprocating cutter bar

NEW DEVELOPMENTS

A current project on the harvesting of grass silage being conducted at the AFRC Institute of Engineering Research (IER) includes work to develop a novel pick-up. It utilises a simple high-speed rotor which has plastic paddles which collect, elevate and convey the crop under a covering hood. Currently, this device is attached to a self-loading forage wagon and, when collecting short grass for zero-grazing, losses were reduced from 25% to 3% compared to a conventional pick-up. It is envisaged that, when fully developed, this device will have applications on big balers and forage harvesters. With respect to trailed forage harvesters, wide (2.5 to 3 metre) pick-ups of simple construction will be possible, thus facilitating the harvesting of wide swaths, which are known to maximise the drying rate of conditioned grass.

This device may also prove to be advantageous when harvesting whole-crop cereal silage, since its sweeping and sucking action may enable more of the grain fraction to be retained. It will not, of course, be capable of recovering grain which has previously been lost amongst the stubble during mowing since, if the suction effect were great enough to achieve this, then soil contamination would also occur.

In addition to the advantage of improved collection of the smaller crop particles, the device also has the advantage that it can form the basis of a simple identification/rejection system for foreign objects, both metal and stone. Experimental results suggest that the continuous stream of crop delivered by this pick-up might have the effect of smoothing out the often excessive fluctuations in forage harvester loading. In addition, chemical additives can be easily applied to the fast-moving crop stream.

An adaptation of this pick-up, specifically for whole-crop cereals, would include a conventional reciprocating cutterbar positioned in front of the pick-up. This would facilitate direct harvesting of the crop and thus eliminate the inevitable grain losses associated with mowing and swathing. It would be quite possible for the cutterbar to be an auxiliary attachment which could be easily lifted out of the way, or completely removed, for operation in grass swaths.

This form of pick-up, although not currently available from forage harvester manufacturers, would enable the farmer to use conventional trailed equipment for harvesting grass and whole-crop cereal silage without changing headers. It would give distinct advantages in both operations. The device is in the later stages of development and there is interest in its commercial uptake.

CONCLUSIONS

Whole-crop harvesting systems for forage are most likely to be based on existing grass silage equipment. The differences between these crops will require particular attention to:
* The cutting and pick-up mechanism;
* The avoidance of grain loss and separation;
* The handling and application of chemicals in dry form, at relatively high rates.

REFERENCES

ADAS (1982) *Combine Harvester Survey 1982*. Internal ADAS communication.

ADAS (1983) *Forage Harvesting Systems Survey 1983*. Booklet BL5035. ADAS, Wrest Park.

BOYD, J.E.L. and LONGMUIR, A. (1983) Whole-crop harvesting of cereals in Scotland. *Agricultural Engineer*, Autumn 1983, 91-93.

WILKINSON, J. M. (1981) *Whole-crop Cereals*. Paper presented at ADAS Mechanisation Officers' Annual Conference, 1981.

DISCUSSION

When harvesting whole-crop cereals it is all too easy to thresh the grain unintentionally and to allow it to separate from the straw.

A method of harvesting whole-crop cereals which has been carried out successfully on the farm comprises the use of a self-propelled Claas forage harvester with a combine header, a trailed hopper for the urea and a pneumatic blower driven by a petrol engine mounted below. Using this equipment, one farmer has harvested his own crop and 160 hectares (400 acres) for other farmers, with no complaints. Milk quality had increased following the inclusion of whole-crop cereals in the ration.

Fertiliser and feed grade urea differ in several respects. There are tighter specifications on feed grade prills and on the chemical purity of feed grade urea. Fertiliser grade urea is generally treated with formaldehyde.

The main problem with applying urea is the quantity involved. Applying urea evenly as a liquid would be easier than as a solid, but the volume of solution required is a major drawback and means that this form of application is only suitable for static, not moving, equipment. A significant problem with using feed grade urea is that it is not available in 0.5 tonne bags and filling the urea applicator is therefore very slow.

The rotor pick-up currently being developed at IER is similar to a grain stripper and was initially designed for picking up small particles of grass. It can, however, be used for whole-crop cereals and the reciprocating cutterbar reduced losses on zero grazing from

25% to 3%. It is 2.5 metre wide and has the advantage of a wide pick-up. The rotor works relatively fast with grass, but is slower for whole-crop material. Separated grain is retained in the system, but it is important to avoid picking up soil when using the device. A manufacturer is currently interested in developing this equipment further.

Neither Mr Rea (Chapter 9) nor Mr Goddard (Chapter 10) used re-cut screens when harvesting. They have been used by others, but care must be taken not to crack all of the grain, particularly if the material is too dry. Re-cut screens use more power when harvesting.

The main effect of urea treatment is to improve the digestibility of the straw component. It has little effect on grain husks, there being very small differences in digestibility between untreated and urea-treated grain. Some physical cracking of the grain may be advantageous.

CHAPTER 9

RECENT DEVELOPMENTS AT KITES NEST FARM

F Rea
Kites Nest Farm, Wotton-under-Edge, Glos, GL12 7PH

SUMMARY

We are now (1991) in our fourth year of using whole-crop wheat on a heavy-land farm with two sealed tower silos. The main effects since the introduction of the whole-crop system have been on the silos and on our margins; the unloaders work with greater reliability and our herd margins have increased substantially. Alkali-treated whole-crop cereal now comprises over 25% of the total forage intake of our cows. The high pH of the whole-crop wheat (pH 8) balances the acidity of our grass silage, and probably accounts for the high total dry matter intakes. Our attempts to achieve the target level of urea addition at harvest have so far met with failure, and improvement in this respect is needed in the future. The introduction of wheat on the farm has meant a reduction in the area of grass. This has resulted in more frequent cutting of higher quality grass crops, which have received higher levels of fertiliser nitrogen (N) per hectare at no extra total cost. In addition to improved milk quality, higher milk yields and improved margins, laminitis has been eliminated. Maize has been grown this year for the first time.

INTRODUCTION

Kites Nest Farm is 102 hectares (252 acres), comprising 50 hectares (122 acres) of grass, 20 hectares (50 acres) of wheat for preserving as the whole crop, and 32 hectares (80 acres) of maize. The farm is on heavy land comprising clay loam topsoil over clay subsoil. Rainfall is 32 inches (80 cm) per year.

The farm carries 136 Holstein cross cows, 35 replacement heifers, and 60 silage beef cattle.

Before whole-crop cereals were introduced, the farm was all-grass, predominantly permanent pasture. Tower silos were erected in 1973 when wilting was fashionable. But the towers require grass of 40 to 50% dry matter (DM) which requires June sunshine, and we cannot produce milk out of first cut silage made in June! In addition, the maintenance cost of the unloaders was unacceptably high and the daily unloading time was considerable. We could not afford to erect clamps to hold 1600 tonnes of silage. Gordon Newman's solution was to put a self-wilting crop, which would unload easily, into the towers.

A 600-tonne earth-walled, hardcore-based clamp has been used for grass silage, instead of the towers, for the last 3 seasons. In spring 1989 one sleeper side was erected, the clamp base was covered with hot rolled asphalt and effluent collection facilities were installed.

THE SYSTEM AT KITES NEST FARM

We are now (autumn 1991) approaching our fifth winter of using whole-crop cereals. The crops have been grown for maximum grain yield, using straw shorteners and normal crop protection treatments. The variety Brock was sown in 1986, Slejpner in 1987 and 1988, and Norman in 1989 and 1990.

Harvest starts at the cheesy stage of grain maturity, and we aim for a DM content between 50 and 60%. We have put whole-crop on top of grass in the clamp, removing the sheet first. The whole-crop material requires consolidation, and good sheeting is essential immediately following ensiling.

In our first year we swathed the crop with a disc mower and discovered it would only feed through the auger to the feed rolls heads first. A significant amount of grain was lost in the process. The crops are now direct-cut using a John Deere trailed forage harvester with an adapter plate which replaces the windrow pick-up, to which a combine header is bolted on. It is better to use a combine header or the crop is threshed at the time of pick-up.

Urea is added by hand to each load after being tipped, before it is buckraked into the clamp or dump box. The target rate of addition has varied from year to year, being 4% of the crop DM in 1987, 1989 and 1990, and 3% in 1988. At the 4% rate, 8 three-gallon buckets of urea were applied per load but, as shown in Table 9.1, the actual addition rate achieved differed from the target. As the ammonia loss from sealed towers is negligible, we are aiming at a 2% inclusion rate this year.

Plate 1 Harvesting whole-crop winter wheat with a trailed John Deere forage harvester fitted with a John Deere combine header.

Plate 2 Front view of the John Deere forage harvester and combine header.

There is no point in wetting the crop if it is harvested too dry; the water simply runs off the material and is not absorbed into it. The presence of water, however, is imperative for the hydrolysis of urea to release ammonia, as the following equation shows:

$$CO(NH_2)_2 + H_2O \rightarrow CO_2 + 2NH_3$$
urea   water   carbon   ammonia
                                   dioxide

Plate 3   Side view of the John Deere forage harvester showing the adaptor plate (centre) for fitting the combine header to the harvester.

RESULTS

Whole-crop wheat analysis

The average analyses of the last four crops of urea-treated whole-crop wheat are shown in Table 9.1, together with the calculated amount of urea added, and the proportion hydrolysed to ammonia.

Table 9.1 Composition and nutritive value of whole-crop wheat 1987-1990

| | Crop harvested in: | | | |
	1987 Wind-rowed	1988 Direct-cut	1989 Direct-cut	1990 Direct-cut
Approximate level of urea added[1] (% DM) (target level)	2.3(4)	6(3)	5.5(4)	8(4)
DM (%)	45.7	48.6	74.7	49.2
pH	6.3	8.3	8.2	7.8
Ammonia-nitrogen (NH_3-N as % total N)	36.0	26.0	8.0	26.0
Crude protein (% DM)	15.5	26.8	24.7	31.8
Ash (% DM)	7.2	6.0	3.9	5.6
NDF (% DM)	–	36.8	36.1	47.0
Starch (% DM)	–	26.9	28.3	14.3[2]
Estimated D-value (%)	65	66	69	68
Estimated ME (MJ/kg DM) (range)	10.4 (9.7-11.0)	10.5 (9.3-11.0)	11.0	10.8
Urea-N as NH_3-N (%)	86	39	13	36

[1] Assuming 9% crude protein in the untreated crop DM.
[2] Starch + sugar.

Notable features of the analyses include:

i) The variable level of urea added, and our inability to get the level correct in relation to the target.

ii) The relatively low pH and high proportion of urea present as ammonia in the 1987 crop, which was the wettest, and which received the lowest amount of urea.

iii) Elevated ash contents in both the 1987 windrowed crop, and in 1988 when the crop was rained on.

iv) The relatively low neutral detergent fibre (NDF) content of the material, compared to a value typically greater than 50.5% of the DM in grass silage. This means that it is inadvisable to chop the crop short, or to allow extensive fermentation to occur, if the objective is to have long fibre, or to maintain high rumen pH values in order to stimulate milk fat.

v) The high pH values of the crops ensiled in 1988, 1989 and 1990.

vi) The similarity in starch contents between 1988 and 1989, despite large differences in DM content. In a dry year, like

1989, the crop can lose water surprisingly quickly, at up to 2.5% per day, but there is apparently little increase in starch content during grain ripening.

vii) The variability in metabolisable energy (ME) content between samples. This probably reflects separation of grain from straw when blown into the towers. The material in the middle of the towers has a much higher grain content than that towards the outside.

We noticed undigested grain in the dung of cows given the 1989 crop, which was very mature at harvest. We were advised to treat the material with caustic soda to improve grain digestibility. This had the effect of liberating ammonia from unhydrolysed urea according to the equation:

$$CO(NH_2)_2 + 2NaOH \rightarrow Na_2CO_3 + 2NH_3$$

$CO(NH_2)_2$ +	$2NaOH$	$\rightarrow$	Na_2CO_3 +	$2NH_3$
urea	caustic soda		sodium carbonate	ammonia

It also helped the problem.

Plate 4 Aerial view of a bunker silo being filled with whole-crop wheat. Urea is being applied by hand.

Table 9.2 Complete diets for cows giving 35 litres milk per day

	Kg per cow per day	
	Fresh weight	Dry matter
Grass silage	8.13	2.08
Maize silage	31.6	11.72
Whole-crop wheat	8.13	4.7
Soyabean meal	2.0	1.8
Rapeseed meal	2.0	1.8
Molaferm	3.0	2.2
Fishmeal	0.5	0.45
High Phos mineral	0.14	0.14
TOTAL	**55.5**	**24.9**

Table 9.3 Milk quality pre- and post-introduction of whole-crop wheat

	1986	1987[1]	1988	1989	1990	1991[2]
Milk fat (%)	4.06	4.25	4.28	4.24	4.11	4.22
Milk protein (%)	3.36	3.43	3.42	3.35	3.41	3.50

[1] Started to feed whole-crop cereals in September 1987.
[2] January to June.

Feeding management

The complete diet for freshly-calved cows giving 35 litres of milk is shown in Table 9.2.

We have found that whole-crop cereals must not form more than 25 to 30% of the forage DM part of the ration for lactating cows. We also feed some molasses and a source of high quality protein.

The beef cattle receive a mix of 90% whole-crop wheat, 9% molasses and 1% fishmeal on a fresh weight basis. Target liveweight gain is 0.8 kg per head per day. Next winter whole-crop wheat and

Table 9.4 Dairy herd margins pre- and post-introduction of whole-crop wheat

	12-month rolling average to:		
	April 1987	April 1989	April 1991
Number of cows	149	136	124
Yield (litres/cow)	5478	5900	6370
Milk fat (%)	4.03	4.33	4.21
Milk protein (%)	3.34	3.41	3.46
Milk lactose (%)	4.68	4.55	4.60
Concentrates			
(tonne/cow)	1.79	1.01	1.32
(kg/litre)	0.33	0.17	0.21
Fertiliser N			
(kg/hectare)	310	278	220
Stocking rate[1]			
(livestock units/hectare)	2.1	1.7	2.15
Margin over purchased feed			
(£/cow)	630	944	1056
(pence/litre)	11.5	16.0	16.6
Margin over feed and fertiliser			
(£/cow)	576	888	1002
(£/hectare)	1213	1556	1878

[1] Temporary reduction in stocking rate after 1987 for farm management reasons; stocking rates are now increasing again.

maize silage will be fed in a 50:50 mix.

The beef cattle did not appear to be growing as well on the 1989 wheat as they did on the 1988 and 1987 crops. This is probably because in 1989 the wheat was too dry when ensiled.

Table 9.3 shows how milk composition has improved since 1986, before whole-crop cereals were fed. However, milk quality during the 1989/90 winter was poorer than in the previous year, and this coincided with the whole-crop cereals being very dry.

DAIRY HERD MARGINS

The dairy herd margins before (April 1987) and after (April 1989 and April 1991) the introduction of whole-crop cereals are given in Table 9.4. Spraying costs for the wheat were £181 per hectare (£72.50 per acre) for the 1987/88 crop, £210.50 per hectare (£84 per acre) for the 1988/89 crop, £251.50 per hectare (£101 per acre) in 1989/90 and £135.50 per hectare (£55 per acre) in 1990/91. The lower growing costs for the whole-crop wheat in 1990/91 are in line with expected expenditure, compared to earlier years.

CONCLUSIONS

The main benefits of whole-crop cereals on our farm are as follows:
* Improved margins;
* Laminitis eliminated;
* Improved milk quality;
* Higher feed intakes – high pH wheat mixes well with low pH grass silage;
* No effluent from silos;
* Less wear and tear on silo unloaders;
* Greater yield of whole-crop cereals per hectare – the DM yield per hectare of wheat is greater than that of grass or maize. Short term grass leys have replaced permanent pasture;
* Reduced area of grass enables more frequent cutting and higher quality grass silage;
* Reduced grass silage means fewer tractors and lower labour costs for silage making – we now have two less tractors on the farm than five years ago;
* Wheat stubble is a good place for slurry spreading;
* Wheat can be grown where maize may be too risky.

CHAPTER 10

WHOLE-CROP CEREALS: A NEWCOMER'S EXPERIENCES

R L Goddard

South Barham Farm, Canterbury, Kent, CT4 6LJ

SUMMARY

Whole-crop wheat silage was made for the first time in 1989. The crop was harvested direct by a self-propelled forage harvester and treated with urea at 25kg per tonne fresh weight. An area of 12 hectares was harvested, and yielded 300 tonnes of silage – a dry matter (DM) yield of 12.3 tonnes per hectare. The silage has been included in the complete diet with apparently beneficial effects on milk quality. In particular, milk protein has increased markedly. The improvement in milk quality is estimated to be worth 1 pence per litre. The introduction of whole-crop silage means that we no longer have the daily chore of chopping straw through the forage harvester.

INTRODUCTION

South Barham Farm is predominantly situated on chalk downland with an annual rainfall of 24 inches. The farm extends to 324 hectares (800 acres), of which 93 hectares (233 acres) are winter wheat. The remaining land is in grass (263 acres), maize (127 acres) and lucerne (180 acres) for dairy cows and followers.

Three hundred summer/autumn-calving cows are milked in two herds. The rolling average milk yield is 6400 litres for both herds. Two Farmhand 360 complete diet feeders, dating back to 1975, are used to deliver feed to one herd of 200 cows, whilst the other herd of 80 cows is self-fed on maize silage, receives an 18% crude protein cake in the parlour, and is bedded on barley straw in yards.

Whole-crop cereal silage was made from 12 hectares (30 acres) of winter wheat for the first time in 1989.

Plate 1   Harvesting whole-crop winter wheat at South Barham Farm, Kent, in July 1989. The self-propelled forage harvester is a New Holland 1880 fitted with a Massey-Ferguson combine harvester cutter.

RECENT DEVELOPMENTS

Following the change to complete diet feeding 16 years ago, the next major development was the installation in 1984 of a 25hp hammer mill for grinding home-produced wheat. Whilst this development reduced feed costs, milk fat content remained a chronic problem, with fat levels typically falling to around 3.5% by December. In 1985 we started to chop barley straw daily through a New Holland 717 forage harvester, to provide a source of long fibre in the diet in addition to that from silage.

The acquisition of a New Holland 1880 self-propelled forage harvester in 1975 enabled wilted grass and lucerne, and maize to be harvested rapidly and chopped short. Recently, the strategy with lucerne has been to mow the crop with a 12-foot wide mower, leaving a stubble of 4 to 5 inches (10 to 12.5cm) and to do no tedding at all to the swath, which is laid on the stubble as wide as possible. The crop is then wilted for 3 to 5 days to a dry matter (DM) content

Plate 2   Filling a silo with urea-treated whole-crop wheat. The silo is consolidated, to prevent heating caused by air pockets.

of about 50% and ensiled without an additive in a clamp silo. The crop is rolled using a 10-tonne crawler to achieve the necessary consolidation.

If the weather stays dry, part of the lucerne is made into hay, again with no tedding at all. The hay is sold as horse feed.

In 1989 a new world record, as follows, was set for lucerne silage quality:

Oven DM (%)	55.2
pH	5.2
Ammonia-nitrogen (NH$_3$-N as % total N)	5.2
Crude protein (% DM)	20.8
Modified acid detergent (MAD) fibre (% DM)	29.0
D-value (%)	68.0
Metabolisable energy (ME) (MJ/kg DM)	10.8

The values are declared on an oven dry matter basis; if corrected for volatiles, they would be slightly higher.

Although this silage is of exceptional quality, its value as a stimulant of milk fat is likely to be less than that of lower quality material of higher fibre content.

In 1989 a change from ground grain to Sodagrain meant that the grain mill became obsolete. The most notable effect of the change has been a substantial reduction in the electricity bill, since the price is fixed by the peak load, which was always achieved when grinding grain. A dust and bird nuisance was also elimated by this change.

Our diet in the winter of 1988/89 for freshly-calved cows, yielding 30 litres of milk, is given in Table 10.1.

Table 10.1 Complete diet for cows yielding 30 litres milk per day: winter 1988/89

| | Kg per cow per day | |
	Fresh weight	Dry matter
Maize silage	15	7
Lucerne silage	13	4
Chopped straw	2	2
Megalac (fat)	0.25	0.2
Rapeseed meal	1	0.8
Molasses	2	1.5
Ground wheat	6	5
TOTAL	**39.25**	**20.5**

Despite high forage intakes and the inclusion of chopped straw, milk quality remained unsatisfactory.

WHOLE-CROP CEREAL SILAGE

Following a 7am phone call from a neighbour urging us to have a go at some whole-crop silage, we adapted the header from a Massey-Ferguson 500 combine harvester to fit on the forage harvester. Essentially, the adaption involved:

i)   Narrowing the exit from the header to match the width of the feed rolls of the forager;

ii) Mounting two brackets with which to bolt the combine header on to the forage harvester;

iii) Altering the drive sprocket so that the header finger bar and bat reel moved as fast as possible;

iv) Adjusting chop length to 2.5 to 4.0 cm (1 to 1.5 inches).

The whole operation was completed in one day.

A 12 hectare (30 acre) area of winter wheat was direct-harvested on 25 and 26 July 1989, leaving a stubble of 4 to 5 inches (10 to 12.5cm). Yield was 10 tonnes fresh weight per acre, equivalent to 12.3 tonnes DM per hectare. Losses at harvest were zero.

On arrival at the silo, each trailer load was spread as flat as possible and 25kg urea per tonne of fresh crop weight was applied through a Vicon fertiliser spreader. The crop was then turned over before being shovelled on to the clamp with a loading shovel. The silo was then rolled with a ten tonne crawler to exclude the air which otherwise causes heating.

Analysis of the urea-treated whole-crop wheat silage indicated that it had a DM content of 52%, a crude protein level of 32% of the DM, and an estimated ME value of 11.2 MJ/kg DM.

Table 10.2   Complete diets: November/December 1989

	Kg fresh weight per cow per day	
	Higher yielders (30 litres)	Lower yielders (20 litres)
Maize silage	15	20
Lucerne silage	13	15
Whole-crop wheat silage	4.5	–
Sodagrain (wheat)	5.5	4
Fishmeal	1	–
Molasses	2.5	1
High phosphorus minerals	112g	112g
Calcined magnesite	28g	28g
TOTAL	**45**	**40**

EFFECTS OF INTRODUCING WHOLE-CROP CEREAL SILAGE

The most notable effect of the change to whole-crop cereal silage (and also to Sodagrain) has been a marked improvement in milk quality. The complete diets given to the cows in November/December 1989 are shown in Table 10.2.

The complete mix fed to the higher yielders was analysed as if it were silage, with the results shown in Table 10.3. It is notable that the pH of the mix was close to neutrality.

Table 10.3 Analysis of complete diet mix for high yielders: November 1989

Dry matter (%)	49.9
pH	6.7
NH_3-N (% total N)	12.2
Crude protein (% DM)	19.0
MAD fibre (% DM)	21.1
D-value (%)	74.0
ME (MJ/kg DM)	11.8

The milk composition of the complete diet and conventional herds on 24 November 1989 is given in Table 10.4.

Table 10.4 Milk composition of the conventional and complete diet fed herds, 24 November 1989

	Conventional herd	Complete diet herd
Fat (%)	3.96	4.11
Protein (%)	3.30	3.63
Lactose (%)	4.62	4.64

The improved milk quality in the complete diet fed herd was worth an extra 1 pence per litre. In January 1992 our average milk protein level was 3.72%.

Other effects of the change to whole-crop cereal silage are that we no longer need to chop straw daily, and we increased the area of whole-crop cereal silage in 1991 to 24 hectares (60 acres).

CONCLUSIONS

The change to whole-crop silage appears to have had a beneficial effect on milk quality, and has eliminated the need to chop straw for the cows. We see whole-crop silage as a user-friendly feed which has reduced harvesting pressure and relieved our grain store, which in a wet year is barely adequate.

DISCUSSION

At South Barham Farm the intention is to continue the use of whole-crop wheat for lactating cows, the present (1991/92) inclusion rate being 8.0 kg fresh weight per cow per day. Any surplus whole-crop cereals will be incorporated into the diets of dry cows and young stock, which will be a useful insurance against potential silage shortages. Whole-crop cereals are now an established and essential ingredient of our complete diet regime and I intend to harvest 24 hectares (60 acres) again in 1992.

Grass or lucerne silage and urea-treated whole-crop cereal silage are entirely different products. The latter does not ferment and should not heat. However, as with grass silage the whole-crop clamp was fully sealed, with undamaged polythene sheeting and a 1-metre overlap. Tyres were used to weight down the sheeting, and the clamp was then netted to prevent bird damage.

Adequate mixing of the urea with the whole-crop is imperative for success. When ensiling the crop, a tractor with a loading shovel was used to spread each trailer load of material from the field, to a depth of 45cm (18 inches) in front of the clamp. Urea was then added to the crop using a Vicon fertiliser spreader. The mixture was then heaped and spread out 3 times, and the complete distribution of urea throughout was checked before the material was pushed onto the clamp. This method of mixing appears to have been extremely effective at South Barham Farm, but great care was taken to ensure that complete mixing occurred, and frequent checks were made of urea distribution.

CHAPTER 11

WHOLE-CROP CEREALS AT HILL FARM

R Sands
Strutt and Parker (Farms) Ltd, Hill Farm, Thorpe Morieux, Bury
St Edmunds, Suffolk, IP30 0NR

SUMMARY

Until last year the 2 dairy herds at Hill Farm were fed a complete diet comprising grass and maize silages, caustic soda- treated straw and wheat, and purchased straights. Although the inclusion of straw was felt essential to maintain good milk fat levels, there were a number of problems and disadvantages with its preparation. Consequently, in 1990, 32 hectares (80 acres) of Galahad wheat was grown for harvesting as the whole crop. The wheat grew well despite very dry weather, but it matured rapidly, and cutting date had to be brought forward suddenly. The harvested crop was treated with 4% urea and ensiled in a clamp constructed with Hesston bales. When fed, the whole-crop silage seemed well-preserved and there were no problems of palatability; milk composition was better than for a number of years. Advice on feeding levels for whole-crop changed from 4 kg per day initially to 5 kg early last year, and subsequently to 7 kg. A similar area of whole-crop will be harvested in 1991 and fed at 7 kg per head per day. However, urea addition will be reduced to 3% of the dry matter (DM), as incorporation was very uniform last year, and efforts will be made to consolidate the crop more to save storage space.

BACKGROUND

Hill Farm is a mixed farm in Suffolk, with 2 herds of cows on complete diets, which until last year comprised grass and maize silages, caustic soda-treated straw, caustic soda-treated wheat and purchased straights. As a high proportion of maize silage was fed it was felt necessary to incorporate chopped straw in the ration to maintain milk fat levels, and to do this a "Stropper" prototype straw chopping machine was purchased. This produced an excellent material as it chopped the straw well and the caustic soda was evenly incorporated. However, we also became excellent engineers as the machine broke down every time it was used. In additon to this, it was invariably found necessary to chop straw at a busy time for land work in the autumn, which entailed drawing a man plus tractor away from autumn cultivations. Added to this, the price of caustic soda has almost trebled in recent years. There was, therefore, a strong incentive to find an alternative feed.

Advice was sought from the farmers who had pioneered the use of whole-crop cereals and we concluded that it was worth attempting to feed whole-crop cereals if a uniform mixture with urea at the desired rate could be achieved.

CONSERVING WHOLE-CROP CEREALS

When Mr Robert Self, our contractor, was approached, he expressd an interest in the concept and as a result he constructed a hopper to trail behind his forage harvester that would blow a metered amount of urea into the outlet spout of the harvester. It was then decided that if whole-crop was going to be a disaster it might as well be done in style, so 32 hectares (80 acres) were assigned specifically to grow cereals for use as the whole crop. This land was growing wheat for the second year in succession, therefore Galahad was chosen as the variety. It was treated identically to the rest of the commercial wheat crop except that no ear fungicides were applied.

In spite of the very dry summer in 1990 the wheat grew well and cutting started at what was estimated to be 50% dry matter (DM). At this stage the leaves were yellowing and the grain was turning cheesy. With the hot windy weather at that time, the crop matured extremely fast and the cutting date suddenly had to be brought forward by one week from the time originally intended.

The chopping progressed very smoothly, with the forage harvester producing a well-chopped sample, and with the urea incorporated evenly at the target rate of 4% on a DM basis.

As there were no silage bays available, a clamp was constructed on a concrete area by forming walls which were 2 Hesston bales high, and lining them with plastic sheets. We soon realised that there had to be a certain amount of compaction to avoid the volumes involved becoming unmanageable. Therefore the whole-crop cereals were pushed up the heap using a tractor and buck rake, but with no concerted effort to consolidate.

FEEDING WHOLE-CROP CEREALS

The results obtained were considered satisfactory in that the whole-crop material produced was a light straw colour and seemed to be well-preserved. There were no problems encountered in feeding it to the cattle and this year the cows have achieved the highest milk fat and protein levels in recent years.

When first making whole-crop cereals the advice given was to feed 4 kg per head, but early last year this was increased to 5 kg per head, and subsequently we were encouraged to feed 7 kg per head. It was significant that cows being fed 7 kg had the highest milk fat levels. This highlights the lack of scientific evidence on which to base feeding recommendations.

THE FUTURE USE OF WHOLE-CROP CEREALS

The results obtained in 1990/91 are considered satisfactory, and this season (1991) we intend to cut a similar area and to grow the crop in the same way. However, as the urea was very evenly incorporated it is intended to reduce the inclusion rate to 3% on a DM basis. Also, with the even urea incorporation and to save storage space, a conscious effort will be made to consolidate the crop. In addition, from the beginning of their lactations all cows will be fed 7 kg whole-crop cereals per head per day.

CHAPTER 12

WHOLE-CROP CEREALS ON A DAIRY/BEEF FARM

M Manning

Kilmore House, Kilmore, Bandon, Co Cork, Eire

SUMMARY

Kilmore House is a mixed dairy/beef farm of 122 hectares (305 acres), of which 103 hectares (257 acres) are grassland. All animals receive a complete diet, which has contained whole-crop wheat for the last two years (1989 and 1990). In spring 1989, 4.4 hectares (11 acres) of wheat was sown and treated as a normal wheat crop grown for grain. At harvesting, a grasshopper mower and trailed harvester were used, and urea was applied by hand. The ensiled material was sealed in a walled pit. In 1990 ten hectares (25 acres) of the variety Alexander were sown in two fields. Seven hectares (18 acres) followed grass, with 3 hectares (8 acres) being late-sown on newly drained and levelled land which had previously grown fodder beet. The crop was again mown with a grasshopper mower, and picked up with a New Holland 2200 silage harvester. Yield was determined with a load cell weigher on an Easi-feeder, and urea was applied with a Vicon spreader at the pit. In future the crop will be direct cut. Whole-crop wheat is included in complete diets fed to the dairy herd, beef cattle and young stock. Milk fat and protein levels have been variable. The whole-crop material was sent for laboratory analysis, which indicated a digestibility of organic matter in the dry matter (DOMD) of about 56%. Growing and harvesting costs were higher for fodder beet and lower for maize compared to whole-crop spring wheat.

BACKGROUND

Kilmore House farm comprises 122 hectares (305 acres), of which 10 hectares (25 acres) are in wheat, 6 hectares (15 acres) grow maize, 3.2 hectares (8 acres) are for fodder beet and 103 hectares (257 acres) are in grass. The farm carries a total of 535 animals, which include

120 cows, 39 dairy replacements, 189 beef cattle, 185 animals up to one year old, and 2 bulls (Charolais and Aberdeen Angus). We have used do-it-yourself artificial insemination for 3 years.

All of the cattle are fed using an Easi-feeder, with no in-parlour feed given. Caustic soda-treated wheat (treated off the combine) has been used and for 2 years whole-crop wheat has been fed.

The farm is divided into 60 hectares (150 acres) at home and 62 hectares (155 acres) some 38 miles away. Silage is drawn home by a haulage contractor.

Our aim is to produce as much feed as possible from our own resources.

INTRODUCTION TO WHOLE-CROP CEREALS

At a Keenan Feed Clinic in March 1989, Gordon Newman suggested growing wheat to utilise as the whole crop. That spring (1989), 4 hectares (11 acres) were sown after fodder beet, which had been grazed by young stock during the winter. The wheat was sown and received the same treatment afterwards as if it were going to be harvested as a grain crop.

Gordon Newman suggested that I should attend an open day on whole-crop cereals at Francis Rea's farm on the 6 June 1989. I was very impressed with what I saw there.

We harvested the 1989 whole-crop wheat using a grasshopper mower to cut it and a Pottinger trailer harvester to pick it up, travelling in the opposite direction of the cut to avoid trashing with the pick-up head. Urea was put on by hand at what was estimated to be the correct rate. The crop was ensiled in a walled pit and sealed using two 500 gauge black plastic sheets, together with a few tyres to keep it from rising. Chop length was approximately 4 inches (10 cm) and the crop yield was 100 tonnes in total.

THE SECOND YEAR OF WHOLE-CROP CEREALS

In 1990 the variety Alexandra was used. Ten hectares (25 acres) were sown, 7 hectares (18 acres) on grass ground and 3 hectares (7 acres) after fodder beet. The 7 hectares yielded 30 tonnes per hectare (12 tonnes per acre) and the remaining area yielded 37.5 tonnes per hectare (15 tonnes per acre). The 3 hectares were sown on the 24

April, which was late as the field had to have drains put in, and some levelling by bulldozer was needed. The crop was sown at 312.5 kg per hectare (125 kg per acre) as it was sown late and no tillering would take place. It received no phosphorus or potassium fertiliser, only nitrogen (N), with 85 units used in total. This is a very fertile field as slurry is plentiful on the farm and as much as possible is put on cultivated ground rather than on grass. The 7 hectares received 3.5 fifty kg bags of 14:7:14 fertiliser and 120 units of N. The crop was again mown using a grasshopper mower and picked up with a New Holland 2200 silage harvester.

Crop yield was estimated beforehand by cutting 1 square metre to calculate how much urea was needed. On the day of harvesting one swath was picked up from the length and width of the field and chopped into a Keenan Easi-Feeder which had a load cell weigher, to determine as accurately as possible the exact yield. The urea was added using a Vicon spreader and a Ford 7610 four wheel drive tractor on the pit. This worked very well.

The 7 hectares were cut when the grain was at the cheesy stage, but the straw was green. Another time I would give the standing crop 3 more days before cutting. The material was put in a clamp and sealed using sand-bags around the edge.

We have a contractor for the 1991 crop, with a Kemper 10 foot header on a Claas harvester to direct cut the cereals, and we will therefore eliminate grain losses.

FEEDING WHOLE-CROP CEREALS

We feed a complete diet to the dairy herd, beef cattle and calves from 7 weeks old. Yields of milk fat and protein have not been stable, varying quite widely at times.

Laboratory analysis of the whole-crop cereals is given in Table 12.1.

COSTS OF WHOLE-CROP CEREALS

Table 12.2 shows the costs of growing and harvesting whole-crop wheat compared to maize and fodder beet. The lowest costs per hectare were for maize, with whole-crop being cheaper than fodder beet.

Maize yielded 57.5 tonnes per hectare (23 tonnes per acre) and fodder beet 80 tonnes per hectare (32 tonnes per acre).

Table 12.1   Laboratory analysis of the 1990 whole-crop wheat

Dry matter (% fresh weight)	46.7
pH	7.9
Ammonia − N as % total N	30.0
Total crude protein (% DM)	32.3
Neutral detergent fibre (% DM)	54.8
Cellulase DOMD (%)	56.2
Starch + sugars (% DM)	18.3

Table 12.2   Growing and harvesting costs for whole-crop wheat, maize and fodder beet

	Costs (£/hectare)		
	Wheat (spring)	Maize	Fodder beet
Ploughing			
Cultivation	30	45	45
Sowing			
Seed	40	47	33
Fertiliser	40	20	80
Spray	51	20	50
Urea	52	−	−
Slug pellets	−	15	−
Harvesting	40	66	65
TOTAL	253	213	273

CHAPTER 13

WHOLE-CROP CEREALS ON A SCOTTISH ARABLE/DAIRY FARM

N Knox
Auldhame Farm, North Berwick, East Lothian, EH39 5PP

SUMMARY

Whole-crop wheat has been fed to dairy cows for one year (1990/91) at Auldhame Farm, which is a large arable/dairy farm. The dairy herd has an average milk yield of 7500 litres per cow and is very tightly stocked, thus whole-crop cereals are a useful additional forage, particularly in dry years, as a decision on how large an area to cut can be made in July. The crop was cut at just over 50% dry matter (DM), which was determined weekly around the critical time, by laboratory analysis, and the material was treated with urea. Laboratory analysis of the ensiled wheat indicated a DM content of about 55%, pH of 6.5, crude protein level of 15% in the DM, and a metabolisable energy (ME) content of 11.8 MJ per kg DM. The whole-crop wheat was fed as part of a complete diet which also contained grass silage, caustic soda-treated wheat, molasses and a mixture of straights. Cow health and fertility are good and milk quality has improved slightly following the introduction of whole-crop cereals, although milk yields are unchanged.

INTRODUCTION

The farm is an arable/dairy farm of 540 hectares (1350 acres), located on mostly good arable land. The total area includes 260 hectares (650 acres) of wheat, 80 hectares (200 acres) of barley, 50 hectares (125 acres) of potatoes and 40 hectares (100 acres) of oilseed rape, with about 72 hectares (180 acres) of productive grassland. Soil type varies from very light to heavy and annual rainfall is around 50 cm (20 inches), ranging from 32 to 63 cm (13 to 25 inches). Wheat yields can reach an average of 10 tonnes of grain per hectare (4 tonnes per acre). The farm is too far north to grow maize.

The farm carries 160 dairy cows which currently have an average milk yield of 7500 litres per cow, with the aim of increasing to 8000 litres per cow. The dairy herd is tightly stocked at 0.28 hectare (0.7 acre) per cow for grazing and silage. The high yield group has access to an exercise paddock and is offered zero grazed grass for 12 hours and buffer fed a complete diet for 12 hours for two months in spring. The remainder of the year the main forage is silage. A complete diet is fed which contains 30% caustic soda-treated wheat in the concentrate mixture of straights.

GROWING AND HARVESTING WHOLE-CROP CEREALS

In 1990 about 7 hectares (17 acres) of wheat were conserved as the whole crop, at a dry matter (DM) content of 50%. The wheat was treated as a conventional cereal crop for grain until cut by a contractor with a grain swather, and picked up with the farm's forage harvester. In practice, its width of 17 foot was too great, particularly as a good crop of wheat was harvested. Urea was applied by hand (25 kg feed grade urea per tonne fresh material) and the treated crop was stored in a pit formerly used for brewers grains. Approaching the estimated time for harvesting, samples of wheat were cut every 7 days and sent to the advisory services for the determination of DM. Over 3 weeks the DM content increased from 35%, to 42% then 51%, and the analysis was felt to be very cost-effective in allowing cutting date to be predicted.

The ensiled material was covered with a double sheet and tyres, and final fermentation quality was good. The density of the 3.3 metre (11 foot) deep clamp was 3.7 cubic metres per tonne.

FEEDING WHOLE-CROP CEREALS

The urea-treated whole-crop wheat was fed in a complete diet with grass silage, caustic soda-treated wheat, molasses and a home-mix of straights. It was initially included at 3.6 kg DM per cow per day but this was increased to 7 kg per day.

The chemical analysis of the whole-crop wheat is given in Table 13.1.

ADVANTAGES OF WHOLE-CROP CEREALS

Whole-crop cereals are a useful complementary feed, particularly in dry seasons when grass is scarce. As a cereal farmer with livestock,

Table 13.1 Analysis of whole-crop wheat silage

DM (% fresh weight)	54.5
pH	6.5
Ammonia-nitrogen (% total nitrogen)	45.1
Crude protein (% DM)	15.3
Digestible crude protein (% DM)	10.2
Ash (% DM)	5.0
DOMD[1] (%)	73.7
ME[2] (MJ/kg DM)	11.8

[1] Digestibility of organic matter in the DM.
[2] Metabolisable energy.

the decision whether to cut cereals for use as the whole crop for the dairy herd can be made in July, depending upon other forage available. The material has fed well, and grain losses have been very low – possibly even below those with conventional cereal harvesting.

Milk yields have changed little while feeding whole-crop cereals, but milk quality has improved slightly. The cows appear to be healthier, the incidence of lameness has decreased and the calving index (361 days) is excellent.

CHAPTER 14

THE NATIONAL WHOLE-CROP CEREALS SURVEY

R F Weller

Maize Growers' Association,
Church Lane, Shinfield, Reading, RG2 9AQ

SUMMARY

The main aims of the National Whole-Crop Cereals Survey, which was organised by the Whole-Crop Cereals group of the Maize Growers' Association, were to:

1. *Identify current methods used for the production, conservation and feeding of whole-crop cereals.*
2. *Assess the benefits of whole-crop cereals and identify any problems.*
3. *Provide information on areas of work where further research is required.*

The 53 farms in the survey were distributed widely over the country, with 63 crops conserved as whole-crop forage. Most were winter wheat, and preserved using urea or a urea-based product as the additive. The crops were grown on a wide range of soils, and most farms were producing whole-crop cereals for the first time in 1990. Seed rates, nitrogen application rates, and the use of herbicides, pesticides, fungicides and growth regulators were noted. Crops preserved as fermented whole-crop material were cut at 26 to 45% dry matter (DM), with alkali-preserved crops cut at 40 to 67% DM. Estimated yields showed large variations between farms. Most crops were cut 7.5 to 15 cm above ground level, and were mainly direct cut with a modified forager and combine header, direct cut with a conventional header, or cut and swathed. Chop length, usually less than 5 cm, ranged from 1.5 to 31 cm. Most crops were preserved as alkaline whole-crop cereals using urea, with 6 crops preserved with urea plus an enzyme. Twenty crops were ensiled as fermented whole-crop cereals using either an enzyme or inoculant, with no additive applied to one crop. The survey covered methods for applying additives, and the composition and storage of ensiled material. Large variations were reported between crops in chemical composition, with the variations greater for the alkali-preserved material. Most of the whole-crop was fed to dairy cattle, on average at 30% of total forage DM intake, with alkaline whole-crop cereals fed to 36 herds and fermented whole-crop cereals to 6 herds. Alkali-preserved whole-crop cereals were fed to beef cattle on 6 farms, at 2 to 8.4 kg DM per head per day; fermented whole-crop silage was fed on 2 farms at either 5 or 6 kg DM per day or ad libitum. A wide range of supplements were fed at varying rates. In 15 herds introducing alkaline whole-crop cereals into the diet increased both milk fat and protein, but only 5 farms attributed the increases solely to the introduction of whole-crop cereals. The majority of farms found whole-crop cereals cheaper per tonne to produce than grass silage, but for some farms lower crop yields resulted in increased costs per hectare compared to grass silage. For farmers the main advantages of whole-crop cereals were: flexibility with guaranteed yield; a single, early harvest using one machine; earlier slurry spreading and reseeding; no effluent problems; and a stable product. The main disadvantages were the larger area required on some farms for conservation; losses of grain; and the need for even distribution of urea through the clamp to ensure good preservation and to prevent spoilage. Farmers made a number of comments on herd health after introducing whole-crop cereals. Overall, farmers suggested that more research was needed, particularly on clamps for fermented crops, and on feeding strategies to optimise the use of whole-crop cereals.

INTRODUCTION

The National Whole-Crop Cereal Survey was organised by the Whole-Crop Cereals group of the Maize Growers' Association, with the main aims of:

1. Identifying the current methods used for the production, conservation and feeding of whole-crop cereals.
2. Assessing the benefits of whole-crop cereals and identifying problems in growing, harvesting and feeding them.
3. Providing information on areas of work where further research is required.

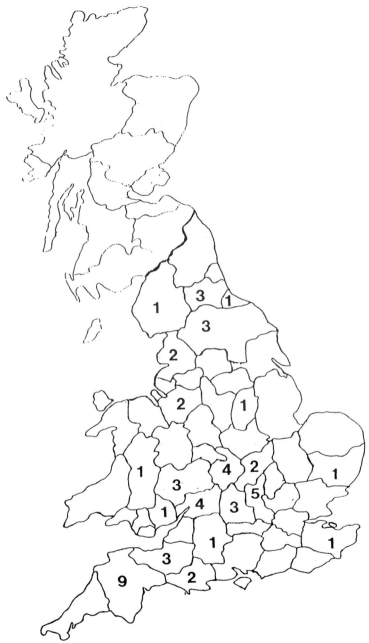

Figure 14.1 Distribution of farms taking part in the National Whole-Crop Cereals Survey, 1990.

As shown in Figure 14.1, the 53 farms in the survey were widely distributed; numbers on the map refer to the number of farms in the county which took part in the survey. The majority of cereals used for whole-crop production were winter wheat crops, with most preserved with alkali, using urea or a urea-based product as the additive.

THE FARMS

The mean size of the farms in the survey was 251.2 hectares, with a range from 44 to 1417 hectares, as shown below:

Farm size (hectares)	Number of farms
<100	13
100 to 200	21
201 to 300	8
301 to 400	3
>400	7

The mean percentage of grassland area on the farms was 58.1%, ranging on individual units from 0 to 100%; one farm purchased a standing cereal crop.

Whole-crop was produced on farms with soils varying from light sand, chalk or gravel to heavy clay, at 8 to 450 metres above sea level. The differences in annual rainfall between farms ranged from 450 to 1379 mm, as shown below:

Rainfall (mm)	Number of farms
<500	3
500 to 750	29
751 to 1000	9
>1000	6

The majority of farms (29) produced whole-crop cereals for the first time in 1990, with 12 farms being in their second year and 4 farms in their fourth year of whole-crop production. The mean area of whole-crop cereals grown per farm was 13.8 hectares, ranging from 2.5 to 56 hectares.

Area of whole-crop cereals (hectares)	Number of farms
<10	22
10 to 20	21
21 to 30	6
>30	3

THE CROPS

On the 53 farms in the survey 63 crops were grown, with winter wheat the main cereal (Table 14.1).

Tables 14.2 to 14.5 provide information on sowing dates, seed rates and fertiliser applications, together with pest and disease control measures.

As shown in Table 14.6, growth regulators were applied to most of the crops.

Crop lodging was reported in 3 crops (winter wheat, winter barley and triticale) despite the use of growth regulators in all crops. As shown in Table 14.7, the majority of crops were harvested between the cheesy dough and hard dough stages.

Table 14.1 Crops used for whole-crop cereals

Crop	Varieties
Winter wheat	Apollo, Avalon, Brock, Carolus, Escorial, Fortress, Galahad, Haven, Hornet, Mercia, Norman, Rendezvous, Riband, Slejpner
Spring wheat	Axona, Sober, Tonic
Winter barley	Gaulois, Igri, Marinka, Torrent
Spring barley	Triumph
Winter oats	Aintree, Image, Peniarth, Solva
Triticale	Lasko

Table 14.2 Sowing dates

Date	Winter wheat	Spring wheat	Number of crops of: Winter barley	Spring barley	Winter oats	Triticale
1989						
August	1	–	–	–	–	–
September	10	–	4	–	2	1
October	24	–	2	–	–	1
November	6	–	–	–	–	–
December	1	–	–	–	–	–
1990						
March	–	3	–	–	–	–
April	–	2	–	–	–	–
June	–	–	–	1	–	–
Range[1]	c. 20/8 to 3/12	9/3 to 18/4	15/9 to 11/10		22/9 to 24/9	30/9 to 5/10

[1]Date within month/month of year.

Table 14.3 Seed rates

Crop	Seed rate (kg per hectare) Mean	Range
Winter wheat	185	156 to 250
Spring wheat	218	156 to 250
Winter barley	178	125 to 210
Spring barley	188	–
Winter oats	180	–
Triticale	179	170 to 187

Table 14.4 Fertiliser applications

Fertiliser applied (kg per hectare)	Winter wheat	Spring wheat	Number of crops of: Winter barley	Spring barley	Winter oats	Triticale
<100	1	2	–	–	–	–
100 to 150	17	1	4	1	–	–
151 to 200	13	–	–	–	–	1
>200	10	2	2	–	2	1

Table 14.5 Herbicide, pesticide and fungicide applications

	Winter wheat	*Spring wheat*	*Number of farms growing:* *Winter barley*	*Spring barley*	*Winter oats*	*Triticale*
Application						
Herbicide						
Spring	11	1	3	1	–	–
Autumn	12	1	3	–	–	2
Both	21	–	–	–	2	–
None	1	2	–	–	1	–
Pesticide + fungicide	23	1	1	–	2	–
Pesticide only	5	1	1	1	–	–
Fungicide only	13	1	2	–	–	1
None	4	2	2	–	1	1

Table 14.6 The application of growth regulators

	Winter wheat	*Spring wheat*	*Number of crops of:* *Winter barley*	*Spring barley*	*Winter oats*	*Triticale*
Growth regulator						
Cycocel	24	–	1	–	2	2
Terpal	3	–	–	–	–	–
Both	5	–	1	–	–	–
None	13	5	4	1	1	–

Large variations were reported in both dry matter (DM) content of the crops and estimated yields (Table 14.8), with crops to be ensiled cut at lower DM contents than those preserved as alkaline material.

HARVESTING

The majority of crops were cut between 7.5 and 15 cm above the

Table 14.7 Assessment by farmers of the grain development stage at harvest

Grain stage	Winter wheat	Spring wheat	Winter barley	Spring barley	Winter oats	Triticale
			Number of crops of:			
Milky dough	5	–	1	–	–	1
Cheesy dough	22	4	3	1	3	1
Cheesy/hard dough	6	–	2	–	–	–
Cheesy dough/ grain hard	1	–	–	–	–	–
Hard dough	9	1	–	–	–	–
Hard dough/ grain hard	1	–	–	–	–	–
Grain hard	1	–	–	–	–	–

Table 14.8 Dry matter content and estimated fresh yields of cereals cut as the whole crop

Crop preservation:	DM (%)		Fresh yield (tonnes/hectare)	
	Fermented	Alkaline	Fermented	Alkaline
Winter wheat	30.0 to 38.0	40.0 to 67.0	12.0 to 38.0	14.5 to 37.0
Spring wheat	45.0	44.0 to 57.0	23.0	14.2 to 30.0
Winter barley	26.0 to 34.0	40.0 to 51.0	20.8 to 37.5	10.0 to 15.0
Spring barley	–	–	6.5	–
Winter oats	28.2 to 34.8	–	15.0 to 30.0	37.5
Triticale	37.2	–	35.0	–

ground, as shown in the following. One crop was cut higher than 15 cm due to the stony ground.

Cutting height (cm)	Winter wheat	Spring wheat	Winter barley	Winter oats	Triticale
<7.5	13	3	1	1	–
7.5 to 15	22	2	3	3	1
>15	4	–	1	–	1

The crops were harvested on the following dates:

	Winter wheat	Spring wheat	Winter barley	Spring barley	Winter oats	Triticale
Fermented material	20/6 to 2/8	20/7	7/6 to 29/6	3/9	14/7	14/7
Alkaline material	15/7 to 20/8	6/8 to 28/8	15/6 to 24/6	–	3/7	–

On 37 farms (69.8%) a contractor was used for harvesting the crops. Although different methods were used for harvesting, 38 farms reported low grain losses during harvest, with high or medium losses reported on 8 farms, where the crops were either cut and swathed (3) or direct cut (5).

Method of harvesting	Number of farms
Cut and swathed	11
Direct cut: modified forager with combine header	27.5
Direct cut: conventional direct cut header	16.5
Other	1

To avoid grain losses through the rear door of the trailers, solid doors were used on 20 farms and mesh plus sheeting doors on 9 farms, with mesh-only doors on 18 other farms.

Differences between farms in chop length were large and values ranged from 1.5 to 31.0 cm:

Chop length (cm)	Number of farms
<5	30
5 to 10	14
>10	3

PRESERVATION AND STORAGE

As shown in Table 14.9, the majority of crops were preserved as alkaline material with urea or a urea-based product; other crops were ensiled as fermented crops with an enzyme, inoculant or no additive. The mean quantity of urea applied was 28 kg per tonne of fresh crop, with application rates ranging from 22 to 50 kg per tonne. Application rates of urea+enzyme+feed (Home'N Dry) ranged from 25 to 35 kg per tonne, with a mean rate of 31 kg.

Table 14.9 Types of additive used for conserving whole-crop cereals

Method of preservation	Additive	Number of crops of:				
		Winter wheat	Spring wheat	Winter barley	Winter oats	Triticale
Alkaline	Urea	35	2	2	–	1
	Urea + enzyme + feed (Home'N Dry)	3	2	1	–	–
Fermented	Enzyme	7	–	3	4	1
	Inoculant	–	1	–	–	–
	None	1	–	–	–	–

Various methods, listed below, were used to apply both the urea and urea-based product to the different whole-crops.

Method	Number of crops
Fertiliser spreader/spinner to apply urea either on a levelled load (prior to filling) or on to the clamp	23
By hand in a dump box	3
By hand over each levelled load prior to clamp filling	3
Using a fore-end loader/industrial loader bucket	3
In the field via the forage harvester	7

Compaction was carried out on 31 crops during clamp filling, with compaction only on the top layers on 2 other farms. Four types of storage were used for the whole-crop cereals, as shown below, with the fermented crops stored in Ag-Bags.

Type of store	Number of crops
Tower	3
Walled clamp (including both soil and straw bale walls)	30
Unwalled clamp	5
Ag-Bag	21

The survey showed large variations in methods of both sheeting and weighting clamps, with some clamps covered with a single 500 gauge plastic sheet plus tyres covering 50% of the clamp surface, while other clamps were covered using side sheets plus a 1000 gauge top sheet plus tyres covering 100% of the clamp surface. The weighting of clamps was either with tyres (30 farms) or straw bales (4 farms). A number of farmers stressed the importance of sealing the clamps correctly to prevent any losses occurring. Surface spoilage or moulding was found in 3 fermented crops. With alkali-preserved materials, 6 crops had surface spoilage, 10 crops showed moulding and 6 crops had both surface spoilage and moulding. The problem of waste was reported as serious in 1 fermented crop and 4 alkali-preserved crops. Of the 22 alkali-preserved crops with spoilage or moulding, 3 were covered with 2 plastic sheets and 5 with both side and top sheets. When clamps were initially opened, 25 farmers reported a smell of ammonia, with no smell reported by 13 other farms.

LABORATORY ANALYSIS OF WHOLE-CROP CEREALS

On 4 farms whole-crop cereal samples from the same clamp were sent to 2 different laboratories. The chemical analyses showed large differences between laboratories in DM content (differences ranged from 0.6 to 3.3%), pH (differences ranged from 0.5 to 2.1), crude protein content (differences ranged from 2.7 to 4.0% of the DM) and metabolisable energy (ME) content (differences from 0.2 to 2.5 MJ/kg DM). These variations may have been due either to differences in analytical methods or to uneven sampling in the clamp. Mean values are shown in Table 14.10.

As shown in Table 14.11, crops ensiled as fermented material had a DM content of 41% or less, with a maximum pH of 4.7. Crude protein content ranged from 8.0 to 12.9% of the DM and digestible crude protein from 4.3 to 8.3% of the DM. Modified acid detergent fibre (MADF) ranged from 13.0 to 34.7%, DOMD *in vitro* from 62.0 to 72.0% and ME from 9.9 to 12.7 MJ per kg DM. The nutritive value of the winter wheat crop ensiled with no additive was similar to that of other fermented crops where an additive had been used.

As shown in Table 14.12, there were large differences in the DM

147

Table 14.12 Chemical analyses of the alkali-preserved whole-crop cereals (mean values with range in brackets)

Additive Crop	Urea Winter wheat	Urea+enzyme+feed Winter wheat	Urea Winter barley	Urea+enzyme+feed Spring wheat	Urea Triticale
Number of samples	30	5	2	4	1
pH	7.6(4.9 to 8.9)	6.3(5.6 to 8.4)	5.7(4.7 to 6.7)	7.6(7.2 to 8.5)	8.1
DM (%)	56.1(45.0 to 83.9)	52.9(44.8 to 60.3)	43.2(40.0 to 46.4)	53.9(43.7 to 62.7)	63.5
Crude protein (% DM)	19.4(11.0 to 34.3)	16.9(15.3 to 17.6)	20.5(16.1 to 24.9)	20.8(17.6 to 24.5)	21.9
Digestible crude protein (% DM)	9.8(6.3 to 12.7)	11.3(10.1 to 12.1)	10.8	12.1(12.0 to 12.2)	–
MADF (% DM)	28.9(15.5 to 39.9)	–	26.5	–	17.4
ADF (% DM)	–	31.1(27.1 to 36.0)	–	33.9	–
NDF (% DM)	45.6(35.1 to 62.7)	58.2(56.5 to 59.8)	45.5	58.8(54.0 to 63.5)	–
Ash (% DM)	5.0(2.8 to 7.9)	–	3.8	4.3	5.2
Starch + sugars (% DM)	19.5(7.5 to 25.6)	–	–	–	–
Starch (% DM)	28.5(24.5 to 33.5)	–	–	16.7	–
Ammonia N (% of total N)	27.7(12.2 to 52.0)	26.8(22.6 to 30.9)	22.1	22.8(11.1 to 31.3)	13.0
Ammonia N (% of crude protein)	9.3(3.2 to 27.8)	–	–	4.3(1.8 to 6.7)	–
DOMD in vitro (%)	67.4(54.4 to 75.0)	68.9(64.8 to 71.9)	70.8(69.9 to 71.7)	66.4(65.8 to 67.0)	–
ME (MJ/kg DM)	10.6(8.1 to 11.6)	12.3(11.7 to 12.9)	11.0(10.8 to 11.2)	11.4(10.4 to 11.9)	–

Note: All samples were not analysed for each chemical component. More samples were analysed for crude protein (36) than for digestible crude protein (16); only 5 samples were analysed for starch, 4 samples for starch+sugars, 5 for ADF, 13 for MADF, 14 for NDF, 14 for ME, 12 for ammonia (% of crude protein) and 20 for ash.

head per day or *ad libitum*.

As shown in Table 14.13, the majority of crops (both preserved with alkali and fermented) were fed to dairy cattle, in 42 different herds. The mean number of cows per head was 177 with a range from 50 to 650 cows.

Table 14.13 Quantities of whole-crop cereals fed to dairy cattle

Method of preservation	Crop (number of farms)	Quantity of whole-crop cereals fed (kg DM/head/day)		% of whole-crop cereals in the total forage	
		Mean	Range	Mean	Range
Alkaline	Winter wheat (30) Spring wheat (2) Winter barley (2)	4.0	1.0 to 8.0	29.8	10.0 to 66.0
Fermented	Winter barley (1) Winter oats (2) Winter wheat (3)	4.7	2.0 to 6.4	30.5	21.0 to 40.0

Further analysis of the data from the dairy herds fed urea- treated whole-crop cereals showed that the percentage of whole-crop in the total forage fed was less than 25% in 8 herds, 25 to 50% in 20 herds and over 50% in one herd. On 38 of the 53 farms in the survey, grass silage was fed with whole-crop cereals to either beef or dairy cattle. Whole-crop cereals were fed with maize silage on 15 farms, with one farm feeding a combination of whole-crop, maize and lucerne silages.

On 9 farms molasses was fed with whole-crop cereals, with concentrate supplements (excluding molasses) for the dairy herds ranging from 1.3 to 12.2 kg per head per day, depending on the type of supplement fed (particularly protein content), stage of lactation and other forages included in the diet. Concentrate supplements included both purchased concentrates and home mixes, the main ingredients of which were soyabean meal, fishmeal, palm kernel meal, oilseed rape, maize gluten feed, cereal grains and sugar beet pulp. Some farms fed a combination of both types of supplement.

Milk yield and composition when whole-crop cereals were fed are shown in Table 14.14.

151

Table 14.14 Yield and composition of milk from dairy herds feeding whole-crop cereals

Method of preservation	Number of herds	Milk yield (litres/cow)		Milk composition (%)			
				Fat		Protein	
		Mean	Range	Mean	Range	Mean	Range
Alkaline	30	6190	5000 to 7500	4.15	3.64 to 4.68	3.39	3.23 to 3.77
Fermented	5	5830	4000 to 6600	4.09	3.80 to 4.44	3.38	3.18 to 3.70

When alkaline whole-crop cereal silage was introduced into dairy cow diets, 15 herds reported increased milk fat and protein contents, 7 herds had either an increase in fat and a decrease in protein content or an increase in protein and a decrease in fat content, with no change in milk composition in one herd and a decrease in another herd. The overall change in 24 herds when alkaline whole-crop cereals were introduced into the diet was an increase in both milk fat (+0.10%) and milk protein (+0.06%) contents. However, only 5 farms attributed increased milk fat and protein levels solely to the introduction of whole-crop cereals into the diet, with the increased milk protein content on 2 other farms also attributed to the introduction of whole-crop cereals.

COSTS OF WHOLE-CROP CEREALS AND THE OUTLOOK FOR 1991

On some farms, although the cost per tonne of whole-crop cereals was lower than the cost of grass silage, the cost per hectare was higher, as the yields of whole-crop cereals were lower than the grass yields, thus increasing the area required for forage conservation.

Costs of whole-crop compared with grass silage	Number of farms
Higher	8
Less	29
Same	8
Depends on the season (eg rainfall, grass yield)	2

Of the 53 farmers participating in the survey, 44 stated their intentions for conserving whole-crop cereals in 1991, with 23 farmers intending to conserve more, 7 the same, 6 less, and the area conserved on 8 farms depending on the weather and grass yields in 1991. Interest from neighbouring farmers in conserving whole-crop cereals in 1991 was reported by 23 farmers.

FARMERS' COMMENTS

Production of whole-crop cereals

Advantages
* Flexible crop (can be either conserved as forage or combined), with a guaranteed yield (especially useful when grass yields are low).
* A single harvest (unlike grass silage) and easier than the combination of combining then baling then drying grain.
* No effluent.
* An earlier harvest than for maize means:
  a) slurry can be applied earlier;
  b) early entry of reseeds (particularly useful on heavy soils);
  c) the calving pattern can be moved as whole-crop cereals are available in the July to September period.
* Low clamp costs - straw bale walls are sufficient.
* Urea as an additive is safer to use than many grass silage additives.
* Alkaline whole-crop is a stable product (1 farm).
* A very convenient and uniform product when made in Ag-Bags (1 farm).

Disadvantages
* Feeding whole-crop cereals requires a larger area for forage conservation on some farms.
* Even spreading of urea can be difficult, resulting in hot spots and pockets of mould in the clamp.
* Harvesting the crop can be difficult if losses are to be avoided.
* Birds can be a problem with open trough feeding and they also cause problems by pecking through the sheeting covering the clamps.
* Storage in Ag-Bags is expensive (1 farm).

* The straw to grain ratio is too high with oats (1 farm) and triticale (1 farm) compared to winter wheat.

Feeding whole-crop cereals

Fermented crops
* Barley is better than oats (1 farm).
* A very palatable feed (1 herd) which improves the quality of the complete diet mix (1 herd) and replaces grass silage on a 1:1 basis.
* Increased intake and cows fitter (1 herd); no effect on intake in another herd.
* Higher milk yields (1 herd), improved milk quality (1 herd), but lower milk quality in 1 herd.
* Fewer digestive problems (1 herd).

Alkaline crops
* Higher feed intakes recorded in 4 herds (+ 10 to 23%), with another farm reporting greater day-to-day variation in intake.
* Increased milk yields (3 herds) and milk quality (7 herds).
* Higher growth rates with beef cattle.
* Health
  a) fewer visits by the vet (4 herds);
  b) improved fertility (2 herds);
  c) less lameness in 12 herds, but increased lameness in 1 other herd.
  d) fewer digestive problems in 1 herd but increased digestive problems in 2 herds when whole-crop cereals either formed over 50% of the forage or when more than 5 kg DM per day was fed.
* A palatable and more consistent feed than grass silage (1 herd), but not as palatable as grass silage plus maize silage (1 herd).
* Less smell than grass silage.
* A useful feed in complete diets, with whole-crop cereals plus grass silage a good combination (2 farms). But without a complete diet feeder it is difficult to mix with other forages (1 farm) and it is not suitable for self-feeding (1 farm).
* Cows are more contented.
* Dung is firmer than on grass silage diets and cows are cleaner.

* Loss of grain both in the dung (1 farm) and left uneaten in the manger (1 farm).

Other comments by individual farmers

* High ammonia levels in the urine.
* High blood urea levels may occur when a high protein supplement is fed.
* On 1 farm cows no longer lick urine – have whole-crop cereals overcome a deficiency?
* A useful feed but what does it replace – silage or cereals?
* More research is required into:
  a) acidic fermented whole-crop cereals stored in clamps, as the cost of storage in Ag-Bags is high.
  b) feeding a combination of urea-preserved wheat with fermented wheat or barley.

Conclusions

Among the questions raised in the survey and the decisions facing farmers are:

* *Crop choice* Winter wheat or another crop?
* *Harvest* What is the optimum cutting height and optimum straw:grain ratio?
* *Type of whole-crop* Alkali-preserved or fermented?
* *Crop DM* What are the optimum DM contents for both alkali-preserved and fermented material?
* *Urea application* What is the best method of applying urea to ensure an even distribution of urea is achieved?
* *Clamp* Should the clamp be rolled or not compacted at all?
* *Wastage* Is spoilage and moulding in the clamp due to uneven urea application, bad clamp sealing or incorrect crop DM levels at harvest?
* *Chemical analysis* How accurate are the present methods of chemical analysis in determining the feeding value of whole-crop cereals for both dairy and beef cattle?
* *Feeding* What is the optimum quantity of whole-crop cereals in dairy cow diets? What is the most suitable type of supplement?

The results from the survey show that large variations occur in the production and utilisation of whole-crop cereals, leading to variations in both the quality of whole-crop material and also the production performance of animals when whole-crop cereals are introduced into their diets.

DISCUSSION

Nine farmers had fed molasses with whole-crop cereals, but the survey had not looked to see if there was a correlation with milk quality. The methods ADAS use to analyse whole-crop material give nutritional information in agreement with the findings of Wye College and Danish researchers. However, some other analytical methods give results which relate less clearly to the practical feeding situation. In the survey, 3 farmers commented that they felt that ME values indicated by analysis overestimated whole-crop cereals compared to cattle performance when the material was fed.

CHAPTER 15

FUTURE DEVELOPMENTS WITH WHOLE-CROP MACHINERY

I Bjurenvall
Biocomb International Ltd, 11 Charles Street,
London, W1X 7HB

SUMMARY

When Biocomb started experimenting with whole-crop cereal harvesting in Sweden 15 years ago, we forecast that it would eventually challenge combine harvesting in many areas. This view is still held. As an engineering R & D company, we were interested in straw as an industrial raw material and a forage harvesting type of operation seemed to be the only feasible approach. Harvesters were initially developed based on a large imported self-propelled forager, before Biocomb produced its own more robust machines which were suitable for a wider range of duties. A pre-production model may be used in the UK in 1991. The company believes that whole-crop cereals have a permanent future as a ruminant forage. However, there are other potential uses for the whole cereal crop. Biocomb has developed machines for a fuel cropping industry which could use whole-crop cereals to generate heat and energy. There may be a change in government policy to encourage fuel-cropping rather than taking land out of crop production, and we support annual crops rather than timber as fuel biomass. The full industrial use of cereal crops implies fractionating the crop into 6 or 7 separate components and Biocomb believes this should be carried out as a farm operation. Small localised plants could then manufacture paper pulp from straw, together with chemical by-products. In the future we believe many cereal growers will harvest whole-crop cereals and either feed them to livestock or fractionate the crop and sell the products to a variety of industries.

INTRODUCTION

Biocomb has been working with whole-crop cereal harvesting for 15 years. In 1977 the company forecast that whole-crop cereals would eventually challenge combine harvesting in many areas of the world, and it sticks to that view. We were then well ahead of our time, but now environmental pressures alone are making the case for us; threshing grain out in the field and leaving all the residues on the ground is not acceptable farming practice – at least not under northern European conditions.

Even if the straw is baled, most of the chaff and other lighter fractions are not picked up. The resultant build-up of weed populations and the risk of pest and disease carry-over in residual crop material commit the grower to increasingly complicated and costly spraying programmes in order to protect following crops. When the straw is uncollected, the chance of disease carry-over is greater. With burning soon to be prohibited, efforts to dispose of straw involve extra energy expenditure on chopping and burial, a lowering of soil nitrogen and possible inhibition of subsequent crop growth. Whole-crop harvesting is cleaner, more hygienic, more energy-efficient, less wasteful and ultimately less expensive.

Biocomb is a research and development company – that is, it sells ideas and knowledge. But the company does not fool itself that using whole-crop cereals is a new idea – it is totally traditional. Until about 40 years ago the farmer always gathered in the grain and straw together, and good use was subsequently made of every part of the crop. Most of the straw was used on the farm, as thatching, insulation, animal fodder and bedding. When converted into manure it was returned to the land to help maintain soil fertility.

Biocomb's starting point was that straw is attractive as an industrial raw material. It's chemical, physical and biological properties make it far too valuable to waste in a world of shrinking natural resources. The problem is that straw is difficult and expensive to handle. But, with the amount of power available today, there must be a way of dealing with straw that makes it easier and cheaper to handle, and therefore more marketable and more profitable for the grower – almost as profitable, perhaps, as the grain!

MACHINERY FOR WHOLE-CROP CEREALS

Developing new machinery is one of Biocomb's main interests, and when considering alternative methods of crop handling, chopping seemed to be essential in order to allow the straw to flow. This implied a forage harvesting type of operation.

Harvesters were developed based on some of the largest self-propelled models available at the time, but straw was found to be a tough material to cut, to gather and to chop – much tougher than grass. So it was decided that Biocomb would design and build its own harvester from basics, with a wide intake, heavy-duty chopping rotor and all-hydrostatic drive, capable of taking crops cut close to the ground with a 6 metre header. At the same time, the company developed a demountable crop container system to make harvesting a single-unit, one-man operation; a crop compression facility to squeeze more of the fluffy crop into a container; and a rotary drum separator to prove that it is mechanically possible to separate almost every grain from the large mass of straw.

By the mid 1980s machinery had been developed to handle whole-crop cereals, but where was the market for the whole-crop material or for the separated, chopped straw? At this time it was becoming evident that the most likely initial use for whole-crop cereals would be as a ruminant feed. Following the dry summers of the 1970s, farmers started feeding chopped, alkali-treated straw. Now, research suggested that a similar but more promising technique, in terms of animal nutrition, might be possible with chopped whole-crop cereals, provided they were harvested at a certain stage of growth. This was put to the test and commercially exploited in the dry 1989 and 1990 seasons, and Biocomb is confident that the success achieved with whole-crop feed as a complement to grass silage in these two years will ensure it a permanent future.

Meanwhile the prototype harvesters, after 10 years of operation, have been retired and are being used as a basis for building updated models by RDM, a leading Netherlands engineering company in Rotterdam. RDM will test new prototypes this year. When the test results have been analysed, the design will be reviewed and a start will be made to explore its market potential.

Biocomb believes that its machine will appeal to contractors, particularly, because it has been designed not merely as a forage and whole-crop harvester but as a tool carrier, or a machine carrier, with a much wider range of uses. It can mount and dismount a combine harvester and work as a conventional combine, if needed. It can also be adapted for baling, muck and slurry application, lime spreading, deep and shallow cultivations and perhaps even root harvesting – all as a single self-propelled unit with considerable load-carrying capacity.

The machine has large wheels, four-wheel drive, a strong chassis, all-hydrostatic transmission and services, and is of all-round robust construction. This is a heavier specification than farmers are accustomed to, and it will not be inexpensive. But this is the way field machines will have to develop when handling raw materials for industry.

Specific industries demand quality, consistency and continuity of supply. For example, forestry machines are tailored for the timber industry, to harvest forest biomass. Now, farm machines for agricultural biomass, which is what whole-crop cereals are, will have to be similar in terms of power, strength, reliability and ease of repair and maintenance.

Biocomb hopes that its machine will soon be harvesting whole-crop cereals in the UK. If all goes according to plan, two models will undergo trials here this summer (1991). One of the first mechanical problems to be solved is the accurate application of urea during harvesting, so as to ensure efficient crop preservation while allowing for a possible reduction in application rate. We shall initially try to carry and apply the urea on the harvester, which is large enough to accommodate a large tank of the chemical.

POTENTIAL DEVELOPMENTS IN WHOLE-CROP CEREALS

Future trends will probably be in the direction of increasing the digestibility of the crop – perhaps mechanically, by excluding a proportion of the coarser stem material, or perhaps biologically, by improving the cow's digestive performance. If the stored crop were to consist of only the grain, the 'chaff and the lighter parts of the straw, this would constitute a more digestible and valuable feed, requiring less in the way of costly supplements to make up complete animal rations. The residual straw could be used for bedding or sold. Alternatively, the whole stem could be left with the crop and it might prove possible to manipulate the appropriate rumen bacteria within the cow so as to enable her to digest it better. Experiments have been conducted along these lines and if such a practice achieved consistent, positive results it could have profound long-term significance for the feeding of roughages to ruminants.

ALTERNATIVE USES FOR WHOLE-CROP CEREALS

One of the potential uses for whole-crop cereals as an industrial crop, outside the conventional food and feed chain, is as a fuel.

Five years ago we proposed a scheme for what was termed fuel cropping, as a means of reducing cereal surpluses. The idea was that the subsidies granted by governments to support the purchase, storage and export of grain might be more profitably employed in diverting part of the national crop into energy generating schemes. Unfortunately, the introduction of set-aside payments cut across that policy by encouraging growers to withdraw land from production altogether. Since that time, however, growing crops for fuel has become a topic of lively interest because of moves towards alternative energy sources and land use diversification.

Regrettably, timber coppicing is the method officially favoured at present. This has been tried in Sweden and rejected for a number of reasons: weeds, pests and diseases, cost and difficulty of spraying, loss of land use for traditional cropping, disruption of land drains, haphazard growth patterns and the need for specialised machines. Annual crops would have the advantage of being grown and harvested by conventional methods. Heavy yields of biomass can be obtained from long-straw wheat or triticale, and even higher yields will probably come from new strains bred in the future.

Alternatively, prolific grasses such as *Spartina* or the bamboo-like *Miscanthus*, or perhaps oil-rich crops with a higher energy value may come into favour.

Biocomb has developed several other machines for use in conjunction with fuel cropping: an unloader-dresser for applying preservatives to bulk crops, a low-energy crop dryer and pelleter, and a range of cyclone burners designed to run on solid biofuels. The increasing cost of set-aside may well dictate a change from non-cropping to fuel cropping – perhaps through a system of differential payments to encourage farmers to grow crops for combustion on their set-aside areas. Such a policy would have far-reaching implications for set-aside, energy generation and use, the environment, cereal surpluses and straw disposal after combine harvesting because any scheme for burning whole-crop biomass would also accept straw.

Fuel cropping, if it happens, will form a perfect bridge between the familiar food/feed uses of cereals and industrial applications of whole-crop cereals. But whether it happens or not, we believe that some form of whole-crop cereal harvesting will be providing raw materials for industry in the not-too-distant future. This development, however, will be accompanied by a further essential technical advance, namely crop fractionation.

In the combine harvester era cereal growers have become accustomed to 2 fractions: grain and straw. Whole-crop fractionation will separate cereals into at least 6 fractions – grain, grain tailings, weed seeds, chaff, cavings and straw stems. When chopped, even the stems might be divisible into nodes and internodes. All these fractions, as their various uses become identified, will command individual markets which will combine to add value to the crop.

Biocomb has its own ideas as to how fractionation might be effected. We believe that it will be possible to do it on the farm, which will enable the grower to benefit economically from using or selling each fraction to the best advantage. As an adjunct to whole-crop harvesting, there may in the future be on-farm drying, fractionation, cleaning, grading, densification, storage, preservation and handling, with no need for large process plants. This would represent an extension of existing grain drying and storage activities, with farms producing a range of products, often on a contract basis, for a range of industries.

One of the obvious uses for straw is as pulp for paper production. Straw itself makes very fine quality paper, or it can be mixed with other pulps to make coarser grades.

New low-energy pulping methods now being developed promise to by-pass the need for giant mills, which are currently considered obligatory to achieve economies of scale, and allow economical, pollution-free production in relatively small, localised plants. An ensured supply of clean, dry, chopped straw would be most acceptable to the operators of such plants, which would probably manufacture a series of chemical by-products in addition to pulp.

WHOLE-CROP CEREALS IN THE FUTURE

In summary, Biocomb believes in the future of whole-crop cereals, which will be fed on an increasing scale, in both fermented and preserved form, to ruminant animals. They may also find a place as a fuel resource, and they will be separated into fractions of increased value for both farm and industrial use.

Outside the prairie areas of the world, Biocomb thinks that all cereal harvesting operations as known today – combining, baling, fields full of long stubble and the disposal of straw as waste – will eventually be superseded to a greater or lesser extent by whole-crop cereal harvesting, which will ensure that every part of the crop is profitably used.

CHAPTER 16

FUTURE PROSPECTS FOR WHOLE-CROP CEREALS IN THE UK

G Newman
The Travellers Rest, Timberscombe, Minehead,
Somerset, TA24 7UK

SUMMARY

With forage maize well-established as an alternative crop on well-drained, low-altitude soils south of the Humber/Mersey line, interest has been stimulated in suitable crops for less-favoured situations, particularly in the north of England, Scotland and Ireland. All the disadvantages of a spring-sown crop, harvested in the autumn, can be avoided by the use of autumn-sown wheat which can be harvested in mid to late July, giving ample time to establish a grass ley or winter rape. A lower input alternative is triticale, with barley and oats giving variable results. As the nutritional quality of the straw is low, ensiling whole-crop cereals by fermentation would not appear to be a logical system of storage, when the crop can be upgraded by the use of alkalis, which also enhance the digestion of the grain. Future developments depend primarily on the European cereal market and the rewards paid to farmers for milk protein and milk fat. Since the intake of cows is improved when whole-crop cereals are fed with grass silage, a significant area of grass could be displaced by cereals for winter feeding. But if, on the other hand, major developments occur in grass silage technology, the need for an alternative will be reduced. Research into whole-crop cereals is near-market, and will no longer be solely funded by the UK government or its agencies. The Maize Growers' Association is sponsoring, along with seed companies and other interested commercial organisations, further experimental work. Limitations on the use of nitrogen and the need to improve its efficiency of utilisation will also favour cereal growing. Whole-crop cereals are self-wilting and are direct-harvested at dry matter contents which do not result in the production of effluent from silos. For these reasons the 200 pioneer farmers who have tried alkali-treated whole-crop cereals in 1990 will probably increase a hundred fold by 1995 as contractors acquire the necessary equipment and researchers fine-tune the technique. Only a radical change in additive technology is likely to expand the use of fermentation techniques, with their lower yield of crop, higher in-store losses and increased costs.

THE PAST

Any speculation on the future of whole-crop cereals should be tempered by a short review of the past, and a reflection of the current 'state of play'.

A 'ways and means' panel was set up in 1978 by the then Agricultural Research Council (ARC) and the Agricultural Development and Advisory Service (ADAS) of the Ministry of Agriculture, to examine whole-crop cereal harvesting, and it reported on the subject twelve years ago. The panel came to some important conclusions and made recommendations for research (ADAS/ARC, 1979), the most apposite of which, in my view, were:

* Maximum yield occurs when the grain is at 35 to 40% moisture content – the hard dough stage.
* The damage caused to grain by cylinder chopping mechanisms is unimportant nutritionally.
* Forage harvesters should cut direct and should be fitted with combine harvester adjustable reels so that lodged areas can be harvested efficiently.
* Combine harvesters discard some 25% of the nutrients in the crop as chaff, all of which is retained in the whole-crop system.
* When treated with sodium hydroxide at 5% of the dry matter, exceptionally high dry matter intakes were achieved. Unfortunately, the material was unstable in the silo and also attractive to rodents, which added to the aerobic losses.

In spite of the latter comment, the report concluded that 'if, by the addition of alkalis to the whole crop at harvest, preservation becomes satisfactory and nutritive value is increased, then the scope for it as a feed for ruminants should be substantially increased'.

'Whole-crop cereal harvesting can provide new opportunities

both for cereal growers and potential cereal growers. The flexibility of the system is such that many varieties could be developed, providing that sufficient information is available from the R & D sector for innovative farmers to establish some of the possible systems available, with a reasonable chance of success'.

The proposed lines of research included:

* The development of effective preservation (having acknowledged the failure of sodium hydroxide).
* The nutritional aspects of combining whole-crop cereals with grass silage in a mixed diet.
* The testing of feeding systems in practice.

The Milk Marketing Board (MMB) Farm Management Services report in 1985 (Jarvis, 1985) concluded that, 'whole-crop cereals provide a good bulk of low energy, low protein silage, usually with good fermentation, which may be suitable for stock with lower requirements than dairy cows'.

The Swedes, in the meantime, had been unsuccessfully attempting to develop their system of whole-crop fractionation. In the same year as the MMB report, a paper by Wilkinson (1985) reminded farmers that, although whole-crop cereals yielded about the same quantity of energy as grass silage, they only required about 30% of the nitrogen fertiliser. Technically, the farming industry had a forage crop which yielded well and cost less than grass, but which had been rejected in practice because of its poor performance.

THE PRESENT

The increasing value of milk protein and milk fat, anxiety to improve carcase quality, greater awareness of inefficient nitrogen utilisation by grass, in addition to increasing dissatisfaction with the performance of animals on grass silage alone, have all contributed to a growing interest in maize and other whole-crop cereals.

Research workers have confirmed the observation of experienced growers, that the benefits of maize silage relate directly to its mature cob content – in other words, the supply of slowly degradable starch in the rumen. The deficiency of protein in maize silage may allow the rumen flora to take advantage of excessive nitrogen in other feeds, such as grass silage, with urea as an extreme example.

As maize silage gives no better performance, in general terms, than the best grass silage, there is no case for attempting to grow the crop in areas known to be climatically marginal. The most successful use of maize silage appears to be in mixed forage diets. Established maize growers in Europe and North America frequently ensile the cobs as ground ear corn. A similar practice is followed with the ears of wheat and barley, particularly in the Po valley in Italy, where double cropping of cereals is possible, if one is harvested at early maturity. The secret of success in this well-established practice appears to be that the straw content is minimised in order to produce a high energy feed.

Whole-crop cereal silage, in other words grain with *all* the straw, is relatively unusual. This confirms that Jarvis was correct in his conclusion, presumably because the fermentation losses incurred in a crop which has a high straw content reduce still further its already low energy content. Additional losses from the potential risk of aerobic deterioration make the introduction of silage bags and enzymes from the USA a welcome, if expensive, innovation to minimise these losses.

Nevertheless, the inclusion of nutritionally improved straw, particularly sodium hydroxide-treated material, in dairy cow rations has increased, both in compound feeds and complete diets, illustrating that the inclusion of an alkaline buffer in otherwise acid, or acidifying diets can be cost-effective. The outstanding success of sodium hydroxide-treated grain, mainly due to the weaker cereal market, also confirms that the need for buffering high starch supplements for acid grass silages is more generally acknowledged and creates the opportunity to use urea hydrolysed to ammonia as an alternative alkali for the treatment of thin-husked grain.

Research has indicated the potential value of the treatment of whole-crops with urea, and about 200 farmers have followed Francis Rea's pioneering on-farm application of research (*exactly* what the Ways and Means Panel Report of 1979 hoped for). Few, if any of them, as far as I know have lost interest in urea-treated whole-crops.

Although we should 'accentuate the positive', it would be unrealistic to 'eliminate the negative'. Therefore before looking ahead, all factors, both favourable and unfavourable, associated with both types of whole-crop cereals should be assessed.

Positive factors of fermented whole-crop silage

* Fermentation can be used for all cereals including barley.
* Earlier harvesting may prevent weeds seeding.
* There is no grain loss.
* There is an opportunity to use effective enzymes when they are developed and proven.
* There is no excess non-protein nitrogen in the diet, so the material balances grass silage well.

Negative factors of fermented whole-crop silage

* The crop is cut before peak dry matter yield (soft dough to soft Brie stage).
* There are fermentation losses.
* Special, narrow silos are required to prevent aerobic spoilage.
* The silage is acidic and contains less starch than material harvested at a more mature stage.
* A short chop length is preferable.
* The silge is attractive to birds and vermin.
* The silage is very low in protein.

Positive factors of urea-treated whole-crop wheat

Nutritional
* Urea is a 'user friendly' chemical, it is not sold in a controlled market and it occurs naturally in the digestive tract.
* The material produced gives improved dry matter intakes and health benefits from its buffering capacity.
* The product does not depress milk quality, because starch degradability is reduced.
* Growth in young stock is improved.
* The product is stable during storage, and in the trough.

Agronomic
* There is no climatic limitation to growing whole-crop cereals.
* The whole-crop product gives the option of early harvesting,

either for winter rape or a grass re-entry.
* Earlier whole-crop harvesting allows the cereal grower to clear headlands, or areas threatened by lodging, contamination or pests, well before combining.
* Slurry can be applied before sowing.
* There is flexibility for the cereal grower to harvest the whole crop or grain according to the availability of other forages.
* The quantity of grass silage made can be reduced, leading hopefully to improved quality.
* Special silage clamps are not required.

Negative factors of urea-treated whole-crop wheat

* There is a relatively narrow 'window' for harvesting – probably two to three weeks. If the crop is too mature, grain is not digested well. If the crop is still 'milky', it ferments into butyric silage.
* If the crop is too dry, no hydrolysis of urea takes place.
* Good mixing is essential - separation could be disastrous!
* The material should be chopped short and the clamp rolled hard to discourage aerobic fermentation.
* Barley and oats are unsuitable for treatment with urea.
* Excessive nitrogen from urea means that molasses or fodderbeet is needed in the diet for efficient utilisation. The new claims for high sugar grass silages could be useful, if these silages help to utilise the non-protein nitrogen from urea-treated crops.
* Whole-crop cereals must be mixed with other forages to avoid selection, which could lead to decreased milk yields.
* More expensive, undegradable protein supplements are needed to compensate for the urea in the whole-crop material.

THE FUTURE

As Eastern Europe changes its agriculture, the cereal market may temporarily strengthen, encouraging growers to combine cereals for cash, but this is unlikely to be sustained in the long term. Further controls on nitrogen use, the need to feed more slowly degradable starch to prevent milk protein depression, and the expansion of contractors who need to extend their harvesting season, will all stimulate interest in whole-crop cereals. As the only specialised equipment required is an applicator and trailer to carry the urea and a combine header adapter, the extra capital involved should soon be recovered.

My recommendations (see Appendix) have been updated after another season (1991/92) of whole-crop harvesting.

RESEARCH

In view of the lack of facilities for near market research, it is important to speculate on where funding could be found for a realistic programme. My suggestions are:

* Commerce will continue to test enzymes, inoculants and chemical agents to improve nutritional quality.
* Commerce will develop ammonia sources which will be released quickly.
* Definition of the correct parameters for reliable alkali treatment is required. This may need farmer-funding.
* More specific dietary recommendations are required, particularly in view of the ammonia-nitrogen present. This may need farmer-funding, but commerce may help.
* Commerce may develop specific varieties for use as whole-crop cereals.

CONCLUSIONS

With forage maize firmly established in many suitable fields south of a Humber/Mersey line, whole-crop cereals will be of greater interest north of that line, and in Ireland and west Wales. The product could also replace sodium hydroxide-treated straw plus some degradable protein and cereal in ruminant rations, so that there is a large potential application in complete diets.

While feeding mixtures of maize and grass silage stimulates forage intake, 25% urea-treated whole-crop material in a maize/grass silage diet appears to increase dry matter intake still further. For this reason I expect several thousand farms to use the technique in 1992. Many of the farmers who have made fermented whole-crop cereal silage this year will change to urea treatment next year.

Whole-crop cereals are no exception to the effect that forward-thinking farmers have on research workers. It was the Maize Growers' Association that took the initiative in co-ordinating reseach and development on the subject by setting up a Whole-Crop Cereal Group, able to fund directly some work and at the same time attracting funding from outside sources, including government and the MMB. The disinterest of the Home Grown Cereals Authority (HGCA) appears to be political.

I shall be surprised not to see 50 000 hectare of urea-treated whole-crop cereals being harvested by 1995!

REFERENCES

ADAS/ARC (1979) *Whole-crop Cereal Harvesting.* ADAS/ARC Ways and Means Panel Report, October 1979.
WILKINSON, J.M. (1985) Best value for money. *Farmers Weekly*, 22 February 1985, 42-45.
JARVIS, P. (1985) *Alternative Forage Crops.* MMB Farm Management Services Report Number **44**, p 64.

APPENDIX

Whole-crop wheat

Maize apart, whole-crop cereal silage has never performed very well. Research carried out in the 1980s at the then Grassland Research Institute at Hurley showed that, when preserved with adequate urea, involving no fermentation, whole-crop wheat has a high intake potential. This has been confirmed by some 200 farmers, when fed at up to 30% of the forage, and this non-fermented, alkaline (pH 8) product appears to have a beneficial effect on rumen function, increasing dry matter intakes and consequently growth rates, health status and milk quality, rather than milk yield. Whole-crop cereals can replace some concentrates and straw in the diet, when mixed with grass silage, and the material is a superb summer buffer feed for grazing cows.

The principle

Urea is hydrolysed to gaseous ammonia which permeates the crop, delignifies cell walls and is toxic to moulds and bacteria; the process takes about 10 days. If the grain is too ripe and the straw too dry (over 60% dry matter), hydrolysis cannot take place and insufficient ammonia is produced to break down the mature husk; therefore large quantities of whole grain are undigested.

If the crop is cut too wet (below 40% dry matter) and it contains soluble carbohydrates, fermentation takes place and, in the presence of ammonia, a foul-smelling, unstable material is produced. If the urea is badly distributed or the heap inadequately covered, aerobic spoilage can take place by moulding.

Total cost per tonne of dry matter is slightly less than for grass silage with a yield of 10 to 17 tonnes of dry matter per hectare.

The crop

Winter wheat and winter triticale give the best yields. They should be grown for high grain yield, including the use of straw shorteners, and the chosen variety should depend on local advice. On a cereal-growing farm, there is greater flexibility as to how much of which crop is cut with a forage harvester, and decisions need not be made before early July. On all-grass farms buying in standing crops of wheat, the decision on how much to purchase may depend on the quantity of first and second cut grass silage made.

Agronomic advantages of whole-crop cereals are that impure stands can be cut, and the chaff which would otherwise be left in the field is fully utilised. Whole-crop cereals also give the advantage of earlier establishment of winter rape and reseeds. On all-grass farms, whole-crop cereals can be inserted into a reseeding programme using a contractor to grow the crop, as for maize.

Fieldwork

The crop is cut when the leaf has changed colour and the field looks yellow, but is unripe (binder stage). The grain should be at the hard Cheddar cheese stage and the dry matter of the whole crop should be 45 to 60%. In hot years, like 1989 and 1990, wheat can lose 1% moisture per day, but more normally moisture loss is about 0.5% per day. The longer the stubble, the higher the quality of the whole-crop material.

Harvesting should be by direct combine header mounted on the forage harvester, with the speed of work about the same as for combining. If a fingerbar or disc mower or rape swather is used, a 10 to 15 cm (4 to 6 inch) stubble should be left and the crop picked up by a precision or double chop machine. A fine chop is preferable as no anaerobic fermentation is involved, but care must be taken to avoid grain loss.

Urea should be applied at 3% of the dry matter, although with precise application 2% is possible. A 10 tonne crop at 50% dry matter will therefore require 360 kg urea per hectare (150 kg urea per acre) or 15 kg urea per tonne of fresh crop. Urea is best applied on the forage harvester, where it can be applied very accurately, or alternatively it can be spread by a fertiliser spinner on tipped loads and mixed well with the harvested material before clamping.

Clamping

As no anaerobic fermentation is involved, only moderate consolidation is required. The crop should be completely wrapped in heavy gauge polythene as quickly as possible. Because no effluent is produced, there is no need to use a silo for storage – hard standing is adequate. The polythene should underlap the sides of the clamp by about a metre to ensure a good seal - the object is to keep ammonia in as well as to exclude air! If the heap is in an exposed position, tyres, for example, should be used to hold down the polythene in the usual way.

Feeding

Urea-treated whole-crop cereal is a very stable product and ideal for buffer feeding in summer. A typical analysis is: dry matter (DM) 50%, metabolisable energy (ME) 10.2 to 10.7 MJ per kg DM, crude protein 18% in the DM, NDF 40% in the DM, and pH 8. If a higher ME value is required, total yield must be sacrificed by cutting to a higher stubble length. In extreme cases, 'ear silage' can be made, which has an ME value similar to that of grain.

Normally whole-crop cereals should not form more than 25% of the forage. Like Sodastraw, they have the effect of raising forage intakes; with grass silage, 3 kg dry matter per day of whole-crop cereals is probably optimal, whilst with maize 4 to 5 kg dry matter per day of whole-crop cereals have been fed successfully. In practice the correct quantity can be found by trial and error as milk yields tend to drop when grass or maize silage of higher ME is substituted by whole-crop cereals. If a higher proportion of whole-crop cereals needs to be fed, molasses or fodderbeet should form part of the diet to provide soluble carbohydrates. The silo face should be left open to allow excess ammonia to escape.

Whole-crop cereals should be mixed with other forages in the same way as Sodastraw, to prevent selection. High yielding cows with acidosis will tend to select whole-crop cereals and, as a consequence, yield will drop as energy intake is reduced.

Urea-treated whole-crop cereals improve milk quality, and usually both milk protein and fat levels are increased.

Urea-treated material is an excellent buffer when wet acidic silage is fed to sheep and beef cattle. It does not cause the increased water consumption associated with Sodagrain.

Warning

Any form of urea, other than feed grade, should be *checked for purity* before use. Urea in excess can be toxic, therefore it is important to spread it evenly and to ensure that adequate moisture is present for hydrolysis to occur.